BIBLICAL
GENESIS
— VS —
SCIENCE
BIG BANG

WHY THE BIBLE IS CORRECT

DAVID ROSENBERG

BIBLICAL
GENESIS
—— VS ——
SCIENCE
BIG BANG

WHY THE BIBLE IS CORRECT

DAVID ROSENBERG

CITI OF
BOOKS

CITIOFBOOKS, INC.
3736 Eubank NE Suite A1
Albuquerque, NM 87111-3579
www.citiofbooks.com
Hotline: 1 (877) 389-2759
Fax: 1 (505) 930-7244

Ordering Information:

Quantity sales. Special discounts are available on quantity purchases by corporations, associations, and others. For details, contact the publisher at the address above.

Printed in the United States of America.

ISBN-13:   Paperback   979-8-89391-810-6
                     eBook   979-8-89391-811-3

Library of Congress Control Number: 2025917864

Table Of Contents

DEDICATION

To my :
great wife Pnina
and my wonderful daughters
Shuli,
Tamar,
Ilana,
Sharona,
Leah
I also want to thank
physicist Norman Cohen for his kind review of this work

PREFACE

In this book on the big bang and black holes you will find much evidence that the Hebrew Bible is most accurate in describing creation of the early universe. The James Webb telescope has found large black holes with and without galaxies in the early universe. The Hebrew Bible, written about 3500 years ago, describes black hole formation on the second day right after the big bang. Science with their fireballs still have no clue on galaxy formation or where the supermassive black holes came from. Since virtually all physicists are taught and believe in singularities, they have been stuck in infinite gravity and can't explain the big bang. The Biblical solution overthrows more than 60 years of cosmology and nuclear physics work with singularities. Using a limitation on matter per volume, the big bang, early galaxy formation, dark matter, dark energy, lack of antimatter become explainable. Scientists today still have no solutions and if you read this book you will find out why. I have tried to make this book mostly without equations and have saved the rigorous work for the end. My papers are available on the Internet Astrophysics Archives, ArXiv.org. I am a former member of the LSU relativity Group, Loni Hyrel 17 and hope you enjoy this book.

David E. Rosenberg

CHAPTER 1
BACKGROUND

After Albert Einstein published his General Relativity in 1915, Edmund Hubble found the Universe was expanding in 1929. The further away the galaxies were, the faster they were going. Tracing everything back, there was an initial start to the Universe. Using General Relativity, there seemed to be no limit on how small the Universe was when it started. There also was a steady state theory requiring an extra term General Relativity which Einstein had added before he knew the Universe was expanding. The Big Bang theory triumphed when radiation from the Big Bang was found in 1964. The questions remained as to what started the Big Bang? What powered the expansion? What was there before the expansion? How were galaxies made and why they looked much the same? When calculations of light elements (deuterium, tritium, helium, lithium, etc.) were made from hydrogen it was found that only 4-5% of matter was involved. What was the rest of this dark matter at The Big Bang and outside all the galaxies? Why is the geometry of the Universe flat-that is why does the expansion energy exactly match the gravitation? In 1939 Oppenheimer and Snyder had written that further gravitational collapses lead to black holes and a singularity. Nothing was then known that could stop sufficient matter from collapsing to infinite density. Most scientists believe this today and are unwilling to entertain alternate possibilities. Virtually every scientist believes the Universe started as a fireball from something smaller than an atom. Imagine the whole Universe packed into something smaller than an atom. As we will see, there is absolutely no evidence the Universe started at such densities and energies. Yet most cosmologists are still searching for equations to describe such a situation. In 1998, two astronomy groups found by using supernova-star explosions as standard candles, that the Universe's expansion was accelerating. Galaxies were further apart than they should be if the expansion of the Universe was slowing up. This came as a surprise to everyone and is known as dark energy. Cosmologists and astrophysicists have yet to explain any of these phenomena satisfactorily. There have been a million papers changing the basic metric or measurement in spacetime. There are very few papers challenging just the use of mass density and pressure in the stress-energy or momentum energy equations. Occasionally they may use negative pressure to try to get out of black holes and the big bang near singularity density and gravity. I am not sure what negative pressure is on a macroscopic basis. No matter how tightly packed matter is, there is always the presumption that the matter is able to move. Neutrons and protons have a definite size and packing them over 100 trillion (10^{14}) grams/cubic centimeter will not allow them to move. Neutrons and protons are composed of quarks which are about 1000 times smaller. Again squeezing them in densities about 1000 times more would not allow these to move. No cognizance is ever paid to this packing process that matter occupies space. Yet these particles have to move to cause gravity. There are a number of cases where neutron stars and even black holes lose gravity. No one has explained them. However questioning the singularity (infinite density) in a black hole will cause scientists to shun you, almost like you are against their religion.

I got involved in Big Bang work when a Rabbi in Monsey NY about 1981-82 asked me to write a paper with him about the Bible. I was to write about the science part which was basically Biblical Genesis. I had worked in mathematical computer modeling in both rubber and plastic polymer reactions and large chemical plant models in scientific engineering prior to my doctoral degree. I had Bachelor and Master degrees in Chemical Engineering and work with a Doctor degree in biological sciences-involving pattern recognition. Investigating science's view of the Big Bang, I found that scientists fanatically believed in singularities. It was enough that black holes could collapse to infinite density called 'singularities' but they believed the entire Universe started smaller than the size of the head of a pin. I never shared this belief. Coming from engineering, I felt that one had to know the limits of equations even for General Relativity. I learned this from my father who had training at NJIT in engineering at the start of the depression as well as at Brooklyn Polytech and NJIT. They were extrapolating General Relativity from densities in neutron stars where it was more or less confirmed to over 85 orders of magnitude in density to black holes and the big bang. This size would render science's big bang in direct confrontation with Biblical Genesis. They were saying a fireball could make the Universe including highly correlated galaxies, cold dark matter and many other highly ordered phenomena. The Biblical version has a cold shell and hot core providing not only cold dark matter to start, a hot core to expand the parts away causing the Universe expansion and later the formation of a firmament (Bible: 2nd Day). This firmament was only found in the 1990s to be the black hole center of all the galaxies. In the mid 1980s, I bought the classical teaching book on General Relativity called 'Gravitation' by Meisner, Thorne and Wheeler. I read and re-read through a good part of it by 1990-91.

By circumstances, I was down in Lakewood NJ those years so I went to nearby Princeton to register to take courses on General Relativity. The elderly lady who interviewed me was going to place me in the Graduate School until she asked why I wanted courses on General Relativity. I told her I thought the Bible had a better big bang model than science. After that she said to me that I wasn't good for their graduate school but that I could sit in the classes without charge as long as the Professor gave permission. As a result of this, I never again told anyone outside my Orthodox community where I got my Big Bang model. I called Neil Turok to get permission to sit in the General Relativity class and found that Igor Klebanoff had taken over after Turok left. He readily gave me permission. Before starting his class each day, I used to walk over to the physics department to talk to big Jim (P.J.E.) Pebbles, Professor of Astrophysics, what I thought was wrong with General Relativity. Due to my reading, I was way ahead of Professor Klebanoff's class at that time. I talked to Peebles about the problems I thought there were with the stress-energy tensor. This is the collection of matter and energy that causes gravity. I never knew until some years later that Einstein also had reservations about the stress-energy tensor. I visited Peebles before quite a number of classes until one day. That day, I said to Professor Peebles, 'With all the matter in the Universe in one place, wouldn't the big bang have to come from some sort of black hole?' I saw his face turn red, he pointed to the door and he said 'Take your model elsewhere!' I went back to Klebanoff's class learning that certain logical questions you can't ask. Looking back at the incident a few years later, I realized that if the big bang originated as a black hole, this would imply that black holes lose gravitation to start the big bang. It would be also responsible for the dark energy in the Universe where galaxies are further apart than expected. Dark energy will be explained later in the book. I took comfort that by teaching things wrong in my opinion, I wouldn't have someone else claiming my ideas first. Professor Klebanoff did leave us with another General Relativity problem. If the laws of General Relativity are followed, an electron should be a black hole. After leaving Lakewood and Princeton over a year later I continued working on my big bang model according to the Bible. In the late 1990's I began doing phone consults with Rocky Kolb. He along with Mike Turner were the authors of the book 'The Early Universe'. I worked with his help on the big bang synthesis of light elements of deuterium, helium, etc. He had given me the computer program his student wrote on light element synthesis starting from hydrogen. At the same time I registered back in NJIT for courses related to astrophysics, but due to scheduling conflicts with my jobs, I was unable to attend regularly. Since I was in NJIT Physics, I was able to put a number of papers on the Internet Astrophysics Archive today known as 'arXiv.org'. They are still there, also listed toward the end of this book, and are based on highly energetic matter being highly confined and unable to move causing the big bang. The stress-energy tensor has highly energetic or any moving matter will cause enough gravity that no big bang could take place. I had no hard evidence of gravitational losses at the time, but it did fit and solve the big bang problems. Subsequently I did find a number of instances where neutron stars and even black holes have unexplained gravitational losses.

About 1980, a type of scalar named phi (ϕ) was invented by Alan Guth to take away the tremendous gravitational energy of the big bang. It allowed the big bang to form from something very tiny much less than an atom at Planck energy and Plank distances. It started inflating the Universe and would not stop inflating in size until it grew to a size that scientists could handle. A number of times cosmologists and astronomers wrote me that my work posted on the Internet Astrophysics Archives(www.ArXiv.org) was unneeded. They had the inflation model they thought would work. This contrived model had a number of problems such that how it started, how it ended and even any evidence that it existed. There simply was no evidence of left over high energy products such as monopoles that the Universe ever had such energies. They only had to find evidence of this inflation to go from very small with extraordinary temperatures to a very hot fireball. Even if there was evidence of such a scalar, it wouldn't solve many other problems that fireballs had such as highly correlated galaxy formation, cold dark matter, dark energy. More on this later. Their problem was on analysis of the cosmic background radiation left over from the big bang, they never could find any evidence for inflation. One time, they thought they had found it by analyzing the polarization of the waves but this was later found to be dust in the Milky Way, our galaxy. Dark matter searches by the Large Hadron Collider now up to 13 trillion electron volts (TeV) has turned up no evidence of any matter other than normal matter. The same result happened for hundreds of dark matter searches by many other techniques with physicists all over the world. Meanwhile my paper which solved everything from the big bang on, was virtually rejected every time I submitted it to an astrophysics journal.. Because I stated there was no evidence for a singularity, many different physics magazines summarily rejected it. I did get some short termed recognition. I started getting involved with the Cactus group that put general relativity on the computer. Their computer program was a gigantic program doing calculations from black holes and the big bang. Their program was written to avoid the infinities of the non-existent singularities. I showed several people from the computer project what dark matter, dark energy were and several other solutions of a big bang that had not started at such enormous energies. Although they disagreed with using lower energies and doing away with singularies (I had zero gravity as the highly squeezed core matter couldn't move), I did earn 8 months on LSU relativity group 'Loni Hyrel 17' starting June 2017. When I still couldn't get my papers accepted, I was not renewed 6 months later. Some months later, after I read that there still was no evidence of inflation, I decided to write this book. To this day, there is no current scientific theory how gravitation completely was eliminated and the big bang happened to expand, exactly matching expansion energy with gravitation. The James Webb telescope has been finding fully formed galaxies called 'little red dots' as close to the big bang as they can see, about half a billion years. Using a fireball big bang, scientists have no solution to galaxy formation despite hundreds of them working on this for many years.

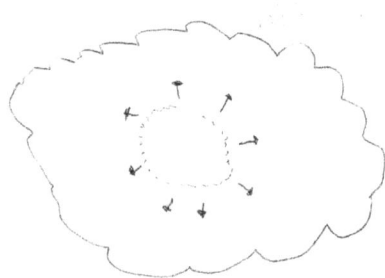

Figure 1: **FIREBALL PROBLEMS:**
1. Fireballs can't match gravitation with expansion energy.
2. Fireballs can't make highly correlated galaxies with rotation, luminosity and black hole mass
3. Fireballs can not explain the cold dark mass in galactic halos.
4. Fireballs with singularities can not explain dark energy.
5. Fireballs with singularities can not explain gravitational loss at the big bang.
6. High energy particles such as anti-matter, monopoles, domain walls are missing.

CHAPTER 2
HEBREW GENESIS TRANSLATION (INCLUDING PHYSICS)

The Bible starts with God creating space and matter. One must realize this was written at times the Hebrews lived in tents around 1500 BCE. Obviously space had to be created first. However the space is described amazingly empty. Any matter over 4 solar masses can collapse to a black hole. Here we have all the matter in the Universe, many billions of solar masses in one place. We will explain collapsing matter by starting with normal matter at a water density of 1 gram per cubic centimeter. There is much space between the molecules. We will crush the molecules into atoms. Each atom has electrons orbiting. There is one electron orbiting for each proton in the nucleus. There are tiny neutrons and protons inside each atom. We will crush the electrons into the protons in the nucleus making them into neutrons with a few protons. Once we have crushed things to about one billionth of a billionth of the original size or one millionth of the length, one millionth of the width, and one millionth of the height, we have something called a neutron superfluid and a lot of space. We have the original space that is now amazingly empty compared to the volume that the matter originally occupied. The neutron superfluid that remains is amazingly heavy. Put it in a bathtub, it would fall to the center of the earth. It has amazingly low viscosity. If you stir honey, the waves die out immediately. Water waves will go back and forth a few times and die out. Stir the bath tub with neutron superfluid, it is so watery that the waves will go back and forth for months. This is the water the Bible is talking about. There was darkness on the surface of the outer shell of this matter. It was not hot, unlike virtually all scientific theories. Most of its energy was transferred to the highly compressed core of the matter. The shell is the cold dark matter that scientists can't find. Due to the fact that the matter is highly squeezed, no particles can move. Once the collapse is complete, that lack of motion means there is no gravity. We have amazingly small super-dense matter in a very large now empty volume without any gravity. God created this from nothing then said 'Let there be light'. The heat and the energy that were concentrated in the core powered the big bang. It definitely was not a singularity of infinite density. There was a large rapid expansion of space and matter. From an open universe, the sudden release of matter and energy was the oppositive of a gravitational collapse. This process was represented by light as there was no air for sound waves to make any noise. This light which has the highest velocity came out first. The darkness was the surrounding space and the cold dark matter shell breaking apart. Time is highly dilated or even stands still at the speed of light or a strong gravitational field. There was no Universal time in our Universe. There was no earth or sun at the beginning for the Hebrew word for day to make any sense. It is logical that the Universe was created in a sequence with some time passing. On the second day, the dark matter collapsed into the black holes of the firmament. These will form the cores of the galaxies. It was over 30 years ago that black holes were found at the center of galaxies and this finally corroborated the Bible. All the parameters of galaxies such as light, size and rotation speed are based on the central black holes which has left the fireball advocates scrambling to this day for an explanation. (See the chapter on galaxy formation.) God divided the waters that were above the black holes from below. The black holes captured in orbits the hot core gases (waters above the firmament) which became the rotating galaxies. The black holes contained the cold dark matter as a highly squeezed neutron superfluid, which were the waters below the firmament. The big black holes in the center of galaxies are millions of solar masses and the whole galaxy rotates around them. After billions of years, the Bible then jumps to our earth. The waters on earth are now regular water composed of hydrogen and oxygen. The primordial hydrogen for each cycle was formed at the big bang as free neutrons broke down to make hydrogen's proton and electron in a little over 10 minutes. Oxygen was produced later by stars which expel matter as they explode as supernova. God had the earth produce grasses and plants, animals in the seas, animals on dry land then man. To this day there is no explanation how life with one celled archaea, bacteria and higher cells with nuclei containing DNA arose. There is no explanation for viruses with DNA or RNA and reverse transcriptase to make DNA arose. The presumption is that viruses somehow escaped from bacteria. The Biblical creation sequence is exactly correct in going from simple grasses to more complicated life forms and there is plenty of room for evolution in between. Scientists taught that the earth was red hot and molten when first created. The Bible states there was a mist that went up from the earth. When zircon rocks were found that formed when the earth did, there was moisture trapped in them. The Bible says man was made from the ground. DNA from Modern Man or some other close relative's DNA may have been used. To make a woman from a man's DNA is even simpler. God took bone marrow DNA from Adam's rib. He replicated the X chromosome and deleted the Y chromosome. Put this nucleus back into a germ cell, it would grow into a woman.

God created the Universe very finely tuned for life. Scientists believe that their fireball created matter and anti-matter in almost equal parts. Only one matter particle in a billion survived and they have no explanation for it, The Bible has highly squeezed matter so that there was no anti-matter possible in conditions well under that of a fireball. If the strong force holding neutrons and protons together were much stronger, hydrogen would be all tied in heavier elements and no water would be possible. If the strong force were much weaker, building heavier elements from deuterium would be impossible and we would have only hydrogen in the Universe. If gravity were much stronger, stars would collapse to neutron star densities and not explode and planets would crush most life forms bigger than bacteria. If gravity was much weaker the galaxies would come apart and planets would leave stellar orbits. Life would be impossible. Neutrons are about 0.2% heavier than protons. If the reverse were true, electrons would combine with the protons leaving only neutron matter. If neutrons were 10% heavier than protons, they would decay into protons and electrons making elements more than hydrogen impossible. With more than 3 space and one time dimension, general relativity is unstable. Carbon can be made from three helium nuclei. If two helium nuclei weren't very sticky to hold and fuse a third nucleus, no carbon for life could form. The scientists call these fine tunings along with others, "the anthropomorphic arguments" to which they have no solution. They have no solution to dark energy, dark matter, why Universe geometry is flat, the imbalance between matter and anti-matter, etc. We know why. God created the Universe as described in the Bible and fine tuned the above examples for life to exist.

The Biblical commentator who was closest to the modern physical understanding of Biblical Genesis was Sforno. The family of Sforno lived and was well known in Italy for generations. There were a number of great Rabbis and scholars with this name but the Biblical commentator was Rabbi Ovadiah Ben Yaakov. Ovadiah was the son of Rabbi Yaakov Sforno and was born about the year !475 in Cesena Italy. He was already a good scholar on the compendium of Jewish Law and commentary on the Bible known as the Talmud as a young child. He also studied mathematics and philosophy. He was nineteen when he went to Rome to study medicine. He was soon widely respected for his intellect. Besides his Jewish friends, a Johannes Reuchlin was a great non-Jewish scholar who wanted to learn more about the Hebrew Bible. Reuchlin spent two years from 1498 to 1500 with Rabbi Ovadiah Ben Yaakov studying the Hebrew Bible and they became good friends. Some years later there were a number of people who wanted to burn Hebrew books as was done in France. It was the intervention of Reuchlin who saved the books from burning in Germany. Rabbi Obadiah wrote a book on the refutation of the ideas and sayings of philosophers who did not believe in God. He wrote it both in Hebrew and Latin and was called 'Ohr Ammim' which meant the Light of the People. He also wrote a book on Euclid's geometry. He wrote his commentary on the Bible called 'Kavonah HaTorah' or thoughts of the Torah (God's Law). His interpretation of the Hebrew Bible follows exactly the Hebrew. He used his great understanding of Hebrew words in presenting their true meaning rather than analogies and mystical interpretations of other commentaries. Due to his wide education and knowledge, Rabbi Sforno was able to predict much of today's scientific understanding of Biblical Genesis. I have translated here some of his Hebrew into modern English.

Bereshit was the beginning of time. This first moment of time was in-divisible as there was no time before it. He (God) created something from nothing that was there prior and this required no time. The Hebrew word for God refers to something eternal. This does not refer to demons or spirits who die like men. He (God) is the source of all visible and non visible things in the Universe. He has provided an existence for them. The Hebrew word 'sham' (there) from 'shamayim' (Heaven) which is distant. The Hebrew syllables symbolize something in the center (us) being equidistant (orbited) from matter (?stars).

CHAPTER 3
1800's SCIENCE LEADING TO SPECIAL RELATIVITY

The science of the 1800's actually began in 1799 when Count Alexandro Volta invented the voltaic pile or first battery. It was composed of silver, damp pasteboard and zinc. He did show his friend Luigi Galvani that it could make frogs' legs twitch. Soon the battery was used elsewhere for electroplating. In 1820 the Danish physicist Hans Christian Oersted was teaching a class about electricity. When he switched on the current, he was amazed to find the needle in a compass left nearby move. He had just shown that electricity and magnetism are inseparable. Every electric circuit causes a surrounding field. Scientists found that the magnetic field around a working electric circuit would cause iron filings and magnets to respond. It was found the strength of the magnetic field was directly related with the flow of electric current. Although no one really understood the phenomena until James Clerk Maxwell almost half a century later, clever inventors soon used the magnetic forces to move physical objects such as to flip a switch, vibrate a diaphragm. Around 1825 Baron Pavel von Schilling, a Russian scientist, invented what was probably the first telegraph. It consisted of a battery for current on a wire and a mechanical device to respond to changes in the current. Few had the foresight to realize it could send messages long distance. Baron spent ten years trying to get anyone in Russia and later Europe interested. When the Russian government finally asked to put a telegraph link between St. Petersburg and Kronshtad in Finland, Baron died 2 months later. However his invention was part of a lesson on electricity by Heidelberg University Professor Muncke. Listening to the lecture was a William Cooke. Cooke had no background in electricity but was visionary in realizing that sending messages long distance was a wave of the future. He came back to England and ordered the telegraph in several parts so that people would not understand his secret. He finally got together a working telegraph and asked Michael Faraday of the Royal Institution for help. Faraday felt the telegraph would work but could not say how far messages could be sent. Cooke then tried to get permission to use his telegraph on the Manchester-Liverpool Railway. This would have been a test of how far the telegraph would work but he was turned down. He was short of money and turned to Charles Wheatstone who was trying something similar after his involvement with the Wheatstone Bridge. Wheatstone did not invent this apparatus which measures an unknown resistance. The Wheatstone Bridge works by placing two sides of a circuit each with resistors and measures the unknown resistance by seeing what balances the two legs. Wheatstone did popularize it after Hunter Christie invented it and the circuit picked up his name instead. Cooke and Wheatstone tried the telegraph apparatus in Wheatstone's lab with four and a half miles of wire without success. They traced the problem to lack of voltage from the battery when using a distance wire. The two changed the system to a galvanometer. A galvanometer is a very delicate highly sensitive system of a needle or a light pen being rotated by current in a coil producing electromagnetism. They took their joint patent out in 1837. They presented their invention to the engineer of the London and Birmingham Railway, a Robert Stephenson. He agreed to sponsor a trial on the Camden Bank. Here was a steep embankment where trains had to be mechanically lifted. The telegraph worked well but was turned down by the Board of Directors despite Stephenson's endorsement. Cooke did have better luck with another railway being built from Paddington Station in London to West Drayton. The builder I.K. Brunel allowed Cooke to set up a trial telegraph on fifteen miles of track. It worked so well that Cooke added six more miles to Slough at his own expense not waiting for approval. Luckily for Cooke, a murder had taken place in Slough by John Tawell. Tawell made his escape by train to Paddington. When he got there the police were waiting for him, having been alerted by Cooke's telegraph. The fact that a murderer was caught using the new instrument caught people's attention. There soon were more railways willing to string the galvanized iron wire from wooden poles. There also were several new patents for new telegraphs. The partnership of Cooke and Wheatstone broke up. Cooke formed The Electric Telegraph Company along with several partners. By 1848 they had 1800 miles of wire and 3 years later it was 11,000 miles of wire. Telegraph lines spread over Europe to Russia and the Far East. In the United States during the 1830's, Samuel Morse and Vail had invented the dots and dashes of the Morse Code. Morse was behind most of the rapidly burgeoning telegraph industry here and set up the first U.S. Telegraph between Washington DC and Baltimore in 1844. By 1861 Western Union had the first working trans-continental U.S. telegraph. Iron wire, strung on wooden poles, worked well in a dry climate. Originally glass or porcelain mounts kept the lines working even in rain. The earth was a sufficient conductor of electricity for the return current. In order to keep things working across lakes, rivers and oceans a good insulator was needed. It had to be flexible, easily molded and reliable and not to age or crack. Eventually gutta percha was found to be the the best insulator. It came from a gum tree in Malaysia and is still use today to fill root canals. The German engineer Werner Siemans found a way to form the gutta percha around a wire in a continuous fashion.

The next task was to lay the gutta percha coated wire on the sea floor between continents. The process was difficult for a ship on the high seas to lay down cable. The first attempt was from Eastern U.S. to the tip of Newfoundland. The Englishman Frederik Gisbourne tried and nearly bankrupted himself. The task was picked up by the paper magnate Cyrus Field. For scientific consultant, a William Thomson of Glasgow University was chosen. After a number of difficulties laying cable from Trinity Bay Newfoundland to Ireland, they finally finished the job after splicing together the link quite a number of places. The cable worked intermittently for about 6 weeks then suddenly completely stopped working. Upon investigating Field and Thomson found that the cable was constructed with many impurities in the copper wire and was frequently way off center in the gutta percha. What seemed to finish the cable was that high voltages used to send some messages. The American Civil War between the states broke out and laying trans-Atlantic cable was put off till the end of the war. By the war's end, better methods of laying cable from the ship's stern were developed. Additionally Thomson had developed a machine that used sound waves to chart the valleys and mountains at the bottom of the sea. Also learning from successful cable laying in the Mediterranean Sea and Persian Gulf during the war, Field tried again to lay cable across the Atlantic. During this second attempt the line broke in a deep area and the grabbing equipment could not haul the end of the cable to the surface. Finally the third attempt took place in July 1866. It was successful as the were no broken cable to repair. Another further successful cable laying used three ships to support the heavy cable and finally there were two trans-Atlantic cables relaying messages. The Atlantic Telegraph Company was in business for real. Cooke, Field, Wheatstone and Morse became famous.

The telegraph became big business with messages sent and recorded on paper tape. Holes were punched for the dots and dashes of the Morse code. Wheatstone had designed a mechanical machine that read the paper tape and sent pulses out on the telegraph. On the other end was a similar machine converting the electric signals back to paper tape. This worked fine on land lines up to 100 words per minute. On the North Sea cable from New Castle England to Frederica Denmark, the dots and dashes became smeared when sending speeds were over 30 words per minute. It was found that the siphon recorder, a more sensitive recorder invented by William Thomson, could do better and handle the 70 words per minute speed on undersea cables. Two fascinating puzzles arose on this cable. Messages from New Castle to Frederica could be sent forty percent faster from than from Frederica to New Castle. People felt that it may have been related to the land line being shorter on the English side. The solution took a few years but Oliver Heaviside solved it. The second was that sometimes faults in the cable improved the received signal. The signal faded but got clearer on the receiving end before failing altogether. Partial faults sharpened the signal transmission and turned smeared signals into sharp dashes and dots. Heaviside supplied the brain power to solve this puzzle also. In doing so, he discovered the way to make the distortion-free long distance lines that we have today. C. F. Tietgen was an ambitious financier with interests in beer, sugar and banking. He was the owner of the

North Sea telegraph line. He formed the Great Northern Telegraph Company and soon had telegraph lines across Russia with the Tzar's consent to the Far East.

Oliver Heaviside was a quiet man usually involved with his own thoughts. He was a telegraph operator who learned most of the science himself. Heaviside was a brilliant man investigating the science of electric transmission as he trouble shooted problems that came up. Heaviside had taught himself calculus and differential equations from available books. He then turned his attention to James Clerk Maxwell's work on electromagnetism. Repairing faults on the North Sea was a difficult process as ship crews had little idea where the cable fault was. They had the pull up the cable by grappling it, cut it and see whether the fault was in one or another direction. Then they re-spliced it and sailed off in whichever direction was the fault. Heaviside went down where the cable entered the sea. He took a battery to the cable and took two current readings. One reading was taken with the far end of the cable short circuited and another when it was open. He then used Ohm's law of voltage equals current times resistance. He didn't know the resistance of the leak but he could use other resistances and voltages to estimate how far away the fault was. Although the repair men didn't believe him originally, they found he was usually correct. When he wasn't close, the resistance of the fault had varied. Heaviside kept a notebook of the nature of his many observations. He had a small side job writing articles for the Electrician. His writings were not clearly written but contained brilliant insights into electrical transmission. He published papers also in the very reputable Philosophical Magazine. Philosophy in those days included physics. His papers caught the attention of William Thomson (later Lord Kelvin) and even James Clerk Maxwell. Thomson who was the most highly respected physicist of the day and famous due to the Atlantic Cable, had high praise for Heaviside.

Heaviside obtained a copy of Maxwell's book on electromagnetic theory. It took him nine years to master the theory. Heaviside already knew about the Wheatstone bridge. He was adept at finding the sixth unknown resistance when the other five resistances were known. He had also been thinking about inventing duplex transmissions, that is sending transmissions in both directions at the same time. He got some wires together and with his brother Arthur found duplex transmissions worked perfectly. He even managed quadrupled transmitting, which is sending two messages at the same time in each direction. Heaviside did have a number of enemies among the Telegraph Engineers of those days. These more senior people were baffled by some of his formulas and didn't believe his duplex work. Oliver's days were kept busy as a telegraph operator for the Great Northern Telegraph company. Business was great and hectic. Oliver used his spare time learning and trouble-shooting problems. One particular puzzle was that land lines had weaker signals in the rain but the transmission rate was unaffected as the signals remained clear. On the undersea link with Denmark, the rain caused the signals to be smeared out despite the signal strength being the same. Since rain couldn't affect the underwater cable, either the 20 miles on the land in England or the 120 miles on the Denmark side was suspect. With much testing and thoughts worthy of Sherlock Holmes, Heaviside tracked down the problem. The problem was the upgraded Wheatstone telegraph transmitter. The complicated system worked well in dry weather on the land and undersea cables. In wet weather it's additional resistance along with the asymmetric land lines and the high capacitance of the undersea cable smeared out the dots and dashes. Heaviside had figured out this most baffling problem. However the administration refused to replace the cause, the Wheatstone transmitters. Another problem was the rate of sending messages was forty percent faster from England to Denmark than in reverse. Also a fault in the cable sometimes improved the quality of transmission. At some places the gutta percha insulation would be frayed down and there was leakage of voltage between the central copper wire and the metal outer sheath. Heaviside analyzed the times signals were clearer and discovered how to make distortion free cables. He worked on how electrical signals traveled down wires. He read books such as Jean Baptiste Fourier's work on the Analytic Theory of Heat nd Maxwell's book on electromagnetism. There were no good references on electrical transmission in wires. Thomson had written some years earlier that the time delay in a received signal is proportional to the product of the cable's resistance and capacitance. Thomson had also written that the maximum rate you could send signals varied inversely with the length of the cable squared. Basic knowledge in those days was that the voltage was equal to the current flow times the resistance or voltage divided by circuit flow equaled resistance. This however was steady state once the transient changes had stopped. Heaviside started with resistance and capacitance of the cable as Thomson did but added in the misunderstood electromagnetic induction. Michael Faraday had discovered electrical inductance but not knowing much math did not supply any equations. He had found in 1831 that voltage could be induced in a circuit by either moving the magnet or moving the circuit. The faster one pushed either, the more the voltage generated. The energy one used pushing or pulling went into the voltage generated. Attaching a battery to a circuit of wire, it took time to build the two energy systems attached to it. The transmission line contained two different kinds of electrical energy. The first was the capacitance as the line stored electrical energy in it. Secondly, the energy went into building up the energy and strength of the magnetic field. After the circuit was broken, it took time for the current to stop flowing due to both the capacitor's charge and the magnetic field energy. Heaviside had to find the right differential equations that gave voltage at any instant anywhere along the line. Heaviside found as done today the first differential equation must be solved for the rates of change of voltage with distance and time. Then he integrated the rates of change to give actual voltages and apply it to the circuit. He may well have been the first person to accomplish this. Thompson had solved a simpler problem as he disregarded the circuit inductance. Heaviside was able to solve the simpler equation using Fourier's infinite series. Thomson was solving a differential equation using a series solution. Heaviside's problem was much harder. He solved the equations with what today is called differential operators. He used the letter p to represent the rate of change with respect to time (d/dt). The normal resistance R in a circuit he used it dynamically as what today is called impedance. The magnetic inductance $V = LdI/dt$ could be represented as Lp. Capacitance which functions as an integral was represented as

C/p. Thus he reduced complicated differential equations to simple algebraic equations similar to LaPlace transforms used today.

Heaviside published his calculations with the title "On the Extra Current" in August 1876 in the highly respected Philosophical Magazine. With Heaviside's trail blazing mathematics, the transmission line became a dynamic active object. Once the operator pressed the sending key down, the voltage and current would often oscillate up and down before remaining at a steady state value. As the circuit was opened, the voltage and current would again oscillate in decreasing amounts until returning back to zero. This oscillation, after the battery was disconnected, he called the extra current. Heaviside checked what would happen if line were suddenly opened at both ends after it was left to 'charge up.' What usually happened was the current would oscillate in the isolated line until the heat generated would dampen the line back to zero. Oscillations happened as the energy went back and forth between the lines with the capacitance holding some charge for its magnetic field. The line resistance put a damper on things producing heat energy which was lost. Additional resistance damped oscillations faster or stopped them before they started. Heaviside published the answer to why signals went faster on the North Sea cable toward Denmark than reverse in the Philosophical Magazine in March 1877. It was entitled "On the Speed of Signaling through Heterogeneous Telegraph Circuits". He also showed how to analyze alternating current (AC) circuits in 1878 almost 20 years before they were introduced. It took much time for Heaviside to realize that the inductance and capacitance working together could make waves. He knew that leaks or faults in cables sometimes made the signal clearer. Finally he wrote this up in an article in the Electrician how to remove the distortion in a transmission line. It was to put a man made fault at half the distance of the line and 1/32 of the resistance of the line. He was close but it took him several more years to realize the one had to take also the capacitance, leakage and inductance of the line into the equation to make distortion free transmission. Using Maxwell's equations Heaviside showed that energy flowed not through metal wires but through the space surrounding them. He also solved the complex puzzle of why East bound signals traveled faster toward

Denmark than back to England. He found the difference in the two land lines was part of the solution. There was only twenty miles on the English side compared to one hundred twenty on the Danish end. Also important was resistances of the battery on one end and the receiver on the other end. Since the resistances were different and well as the cable resistance, signals to Denmark were quicker because the receiver had more resistance than the battery. East bound signals were faster because the receiver had more resistance than the battery. In this direction, these had lower landline resistance and lower terminal resistance. If the resistance of the receiver was less than the battery, that direction would have been quicker. Heaviside published this genius work in the Philosophical Magazine in 1877. Heaviside as usual did not include his reasoning about charging time which would have made his explanation here so much clearer. He founded electric circuit theory but its acceptance would have not taken many years with just a few clear explanations. Instead of complex differential equations, Heaviside was able to write the equations using the differential operator p for rates of change *d/dt* and make algebraic equations. About the same times as these were published, Alexander Graham Bell patented his telephone. It worked as sound waves vibrated a thin metal plate. This varied a magnetic field which varied an electric current toward down the wire to the receiver. Improvements were made by others on the telephone. Improving the wire was most important as human voice varied up to three thousand cycles per second and more. The large telegraph companies were keen to put the telephone into service along their lines. Heaviside's distortion free technique had been written up and published. He had never patented it and never made the money he should have.

James Clerk Maxwell had died prematurely at age 48 in 1879. He had taken nine years to transform Faraday's ideas of lines of magnetic force to mathematical equations. He was able to calculate the speed of light with his equations. Heaviside had studied Maxwell's one thousand page work on electromagnetism. There was no mention of the telegraph but it added to Heaviside's understanding of everything related. Heaviside introduced the mathematics of vector analysis in 'The Electrician' in 1882. It was the math necessary to understand Maxwell and Heaviside made the understanding much clearer. In 1886 he published in the same journal a series of articles on electromagnetic induction and its propagation. He explained that energy flow was greatest when the forces of electric energy and magnetism met at right angles $E \times B$. Heaviside simplified the many symbols for each force in all 3 directions to one vector using bold print. His vectors were unrelated to any specific coordinate system. Much like Einstein a few years later, Heaviside would make a mental picture than commit it to mathematics. The curl of vectors can be thought of using the bathroom shower. The faster the shower water goes, the more the shower curtain approaches at right angles. The swirl or turbulence of water movement like the strength of magnetic flow will cause electric flow perpendicular or vice versa. No magnetic flow is generated in the direction of electric flow. Thus the changes in electric flow will generate magnetic force in a perpendicular direction. A changing magnetic force wraps or curls around an electric force. The changing of an electric force wraps itself around the magnetic force and electromagnetic waves are generated. The frequency of the changes relates to the frequency of the waves. This is true for all kinds of waves from the very high frequency energetic gamma waves down through lower energy waves of ultraviolet light then infrared. Using Maxwell's equations, one can find the speed of light which is an electromagnetic wave. Another term used is divergence often abbreviated div. Divergence is a summation term. If all the flow into a volume is the same as what comes out, the divergence is zero. If something is added to the flow, the divergence will be positive. If something is lost, it will be negative. In empty space, there is no magnetic force or electric force lost and their divergences are zero. This is not the case if some conducting metal or insulation is present in the field. Heaviside took Maxwell's original twenty equations and reduced them to eight. Heaviside explained Maxwell's equations in much simpler form using vectors. Using symmetry he worked out a complete correspondence to just four equations. Since the concept of a monopole was then unknown, he set the magnetic conduction at zero. Here was at last the four Maxwell equations that every physicist and engineer today knows and are much easier to learn. E is the electric field. B is the magnetic field. The curl **E** is proportional to change of **B** with time. In English this means that the flow electrical energy will cause the magnetic field at right angles to change. curl **B** is proportional to change **E** with time. Here it means the flow of the magnetic field will cause current to flow at right angles. *div* **E** = **0,** *div* **B** = **0** and the divergences are sum terms where none of the electric or magnetic energy is lost. However Heaviside's contribution today is unknown. The Maxwell equations are given in a way simplified by Heaviside without due credit. Like gravity, electromagnetism and light are governed by an inverse square law. If one doubles the distance away from the source, their intensity drops to one quarter. The reason is that the sphere from which the waves travel has a surface area of $4\pi r^2$. As the distance is increased the surface are of the sphere is r squared, and the waves are reduced accordingly. The same principle applies to gravity. Heaviside described the flow of energy outside the wires of a circuit into the magnetic and electric fields generated by the flow of electricity. The wires were merely a guide for the fields of energy flowing outside them. Whatever energy did enter the wires went into generating heat. There was electrical current carried by the wires but the energy paralleled the wires on the outside.

Heinrich Hertz was the first person to send and receive electromagnetic waves 1885-1886. He was a Professor of physics at the Karlsruhe Polytechnic Institute. He had been investigating an induction coil. This coil transformed low voltage direct current to alternating current of much higher voltage. The electricity crossed a small gap in the wire by sparking. Changing the equipment, he added a secondary spark gap in the wire on the other side of the coil. He found that the secondary spark gap also sparked except when the length of wire from the coil was the same on either side of the secondary spark gap. Since this was an alternating current system, Hertz figured that electric waves were traveling through both sides of the secondary spark gap. When the circuit lengths were the same, each side had the same waves canceling each other so no spark was generated. Hertz continued to do experiments. He found the main spark gap produced beautiful even waves which he could predict from the resistance, capacitance and inductance of the circuit. This was all based on Maxwell's equations. Hertz next changed the apparatus.

The was a primary circuit with a coil and a spark gap. Separately he arranged a secondary circuit at some distance also with a spark gap. Hertz adjusted the resistance and capacitance of the main circuit so that it produced waves of 100 million cycles per second according to Maxwell's theory. This would translate to a wave length about one meter. The secondary circuit at a short distance was a meter across. It sparked and shook violently when the primary circuit was electrified. With this and a slightly more complex transmitter and using the same wire with a spark gap as a receiver, Hertz definitely established that electromagnetic waves can be transmitted and received. Hertz was the first to show electromagnetic waves travel outside and along the wires. He also was the first to show that the waves could be detected with nothing more than a loop of wire for a receiver. He moved the loop of wire at various distances from the circuit. Some distances there were strong sparks in the loop of wire and other distances there were no sparks. He had detected standing waves with nodes of no sparks and anti-nodes where the sparks were the strongest. He noted the waves could go through thick wood but were reflected by thin or thick metal. Hertz also varied the frequency of the waves by changing the capacitance and inductance of the circuit. He varied the receiving wire to match. Using the frequency and wave length allowed Hertz to calculate their speed which was the speed of light. Hertz also managed to polarize the waves like light was polarized. Most ingeniously he stretched two copper wires three centimeters apart along a giant frame. Like light which was transverse waves, the electromagnetic waves that passed through Hertz's apparatus were all polarized. He wrote up his findings in three pamphlets. Although German Physicists were not interested, the papers caused quite a stir in England. The Maxwell proponents in England were longing for a proof of Maxwell's equations and Hertz supplied proof beyond their wildest dreams. He reduced Maxwell's equations to the same four as Heaviside. However Hertz felt that the wireless waves would not have any practical application. Heaviside had meanwhile perfected his system for distortion free transmission of electrical signals. It was more needed by the telephone which required more electrical voltage than did the telegraph. Heaviside's brother Arthur wanted to put the telephones in parallel rather than series. This was called a bridge system and reduced cross talk on the phone lines due to inductance. Each telephone would then have a small leak of electricity. Heaviside had previously shown that small leaks would sometimes reduce the distortion of the line. So carefully done, each phone would maintain a distortion free transmission. Unfortunately, the small loss of power with each phone would soon enough reduce the voice to inaudible. Resistance in a phone line slowed the flow of current where as leaks increased current flow with small losses with each leak. Heaviside first thought of an ideal transmission line with capacitance and inductance. He then put in resistance like a real cable. He balanced the resistance of each part of the cable with leaks of varying strength. Each resistance held up and reduced current flow. It did reflect part of the voltage wave back up the line. The leaks increased current flow and reduced charge. Properly balanced, the net effect of the two was to slightly reduce the strength of the waves, but conduct waves down the cable without distortion. Originally Heaviside did this in his head with pictures. He then wrote a differential equation including the resistance, capacitance, leakage and inductance. He was startled to find the way to make a distortion free line was to have the inductance divided by the resistance equal to the capacitance divided by the leakage. The resistance and capacitance of the lines were essentially fixed. So the inductance had to be varied for each leakage. Enemies of Heaviside's brilliance held up publication until June 1887 when the paper appeared in The Electrician in 'Electromagnetic Propagation and Its Induction'. It later also appeared in The Philosophical Magazine and in Heaviside's book on electromagnetic theory in 1892. Heaviside wrote further about the electric field generated by the voltage traveling outside the wire. The magnetic field was generated at right angles by the current flow. In December 1888, Heeaviside published a paper in The Electrician that a moving charge near the speed of light will suffer a contraction in the direction of motion by the factor $v/square - root[1 - v^2/c^2]$. This was the precise factor Einstein used in formulating special relativity(that there is contraction changes moving near the speed of light) some 17 years later. Heaviside did not further investigate these properties. Einstein used the speed of light as a constant and worked the other way for length contractions.

In 1893 Heaviside published a paper in The Electrician with the title 'A Gravitational and Electromagnetic Analogy'. Here Heaviside suggested that the four equations he made from Maxwell's theory could form a good theory of gravitation. This would be correct in a weak gravitational field in flat spacetime. This was twenty two years before Einstein's General Relativity. It is believed Einstein was aware of Heaviside's work.

The scientist who contributed most to 1800's science was James Clerk Maxwell. He was born in Scotland in 1831 and was inquisitive from childhood on. He was in the University of Edinburgh from 1847.There he studied mathematics, logic and philosophy. In 1850 he studied advanced mathematics at Trinity College. He began doing research on the properties of the colors of light. After a bout of smallpox, he developed equations for electromagnetic waves between 1860-1862. His equations were based on previous work by Michael Faraday who gave descriptions but was untrained in mathematics. His equations were subsequently simplified to four equations by Oliver Heaviside. Maxwell's contributions included electromagnetic waves travel at the speed of light. He predicted oscillating waves in electric and magnetic fields. He worked on optic and color. He helped explain the color blindness in some people. He contributed to work on the thermodynamics of steam engines.

CHAPTER 4
SPECIAL RELATIVITY

Before 1900, there was no reason to question the validity of Maxwell's equations for electromagnetic waves. The speed of light could be calculated from constants in them to be 299.79 million meters per second. Maxwell's theory predicted oscillating electromagnetic waves through empty space. Light was presumed to be similar to sound waves and wave equations in both were similar. Sound waves caused longitudinal or vertical pulses in the medium as the waves traveled. Light waves traveled with horizontal or transverse pulses. Sound required a medium to transport it and originally it was felt that light should need a medium too. This was called an ether It had to fill empty space yet not be disturbed by material bodies traveling through it. In 1887 Michelson and Morley set out to measure the effects of the ether. In a cleverly designed experiment, a light beam was split into two. It was sent in two directions to measure the effect of the ether on the light beam. Since the earth traveled in an elliptical orbit, it was going one direction at one time and the opposite direction 6 months later. The results of the experiment showed no difference or no effect of the 'ether' no matter which way the earth was going. Shortly thereafter, there was recognized a conflict between Newton's mechanics and Maxwell's electromagnetic descriptions. Albert Einstein worked on electromagnetic waves of light in a vacuum while he worked at the Swiss Patent Office. Newton's laws were valid in constant velocity situations also known as inertial reference frames. Newtonian mechanics allowed an observer to move at any speed if enough acceleration was present. He asked how electromagnetic waves traveling at the speed of light might appear to an observer traveling at the same light speed. It seems the oscillations of light waves would not be visible as the observer would see only one small section of light. Using Newtonian mechanics the wave motion shown by

Maxwell's equations would not be visible. Einstein realized that a moving observer in a closed chamber traveling at the speed of light would not see any light rays. Without reference to outside locations, he could not tell he was moving at all. With these, Einstein made two postulates:

1. The laws of electromagnetism hold with all velocities and accelerations that classical mechanics are valid.
2. The speed of light in a vacuum is constant for all observers regardless of their motion.

The first principle implies that observers moving at constant velocity must agree on physical laws. The second is that these observers must agree that light travels at constant speed regardless of the speed of the observer or even the source. This second law seems to defy logic. A ball thrown forward from a moving train will travel faster than its thrown velocity by the speed of the train. Likewise a ball thrown from behind a moving train will lose velocity relative to the ground by the speed of the train. One must remember that light is not matter and travels as waves the speed of which are not affected by the movement of observers or sources. However different observers moving at different speeds must still measure the same velocity for the speed of light. The old Newtonian mechanics and common sense break down at relativistic speeds near that of light. This implies that things traveling at or near the speed of light will suffer a length contraction in the direction of motion. An observer moving with speed 'u' shines his flashlight at another observer who is stationary. The light waves travel at the speed of light 'c' despite the fact that the first observer is moving. Even if the first is observer is traveling close to the speed of light, the light rays from his flashlight still travel at 'c' the regular speed of light. There is no additive speed contrary to our common every day sense. Let a light flash from the moving observer shine toward the observer standing. The light flash travels less of a distance since the moving observer moves from the time the light flash was sent until the time it was received. This may be in nanoseconds or billionths of a second where light travels about 1 foot. The same principle occurs even if one reverses the picture in the opposite direction. There is no difference as long as one observer is approaching the other.

When light is flashed from a moving observer with velocity 'u' toward a stationary one, there is a lapse of time. If the stationary observer has a mirror, the light will be reflected back to the moving observer with a flashlight. As the moving observer flashes his light toward the stationary observer with a mirror it needs a certain time to reach the mirror. On the return trip from the mirror the distance is shortened as a result of the speed 'u' toward the stationary observer. Since the speed of light is constant and velocity times time is distance, there is time dilation in the direction of travel.

The time is dilated by a factor $1/square\ root[1 - u^2/c^2]$. The number on the bottom is a fraction which becomes smaller as the speed 'u' approaches that of light 'c'. At the speed of light the bottom is zero and time is infinitely dilated. Einstein was imagining one day when the clock struck three in Berne as his train left the station. If he could ride the light wave of the clock at three o'clock, he would never see any change in time. There is a little ditty that physicists say. "Cynthia Bright could travel faster than the speed of light. She left home one day and came back the previous night." The problem is no material object could travel even at the speed of light. Proper time is dilated only when traveling close to the speed of light. An observer in motion will disagree on length in the direction of motion compared to stationary observers. As time is dilated, length is contracted by the same amount. As length and time are changed in same way but opposite directions (time dilated, length contraction) using the same formula above, clocks also run slow at large fractions of the speed of light. Say we are moving at half the speed of light. Using the formula above: One half squared is one quarter. One minus one quarter is three quarters. The square root of three quarters is 0.86. (0.86 times 0.86 is 0.75). One divided by 0.86 is 1.162. So at half the speed of light, time or clocks are dilated by about sixteen percent and length is contracted by sixteen percent only in the direction of motion. Believe it or not, even rulers and measuring sticks are contracted if moving are good fractions the speed of light by the same formula as above. For a sub-atomic particle reaching the earth's atmosphere at 99.9 percent the speed of light (0.999c), the time dilation factor is 22. Thus the particle will age 22 times slower than if not traveling relativistic speeds.

Distances that light can travel in a time interval are called light-like. Distances require light to move faster in the time interval than the speed of light are called space-like and are not possible. During big bang expansion, most of the galaxies are at space-like distances from others which means there is no one signal that can shape them similarly. This is one of sciences puzzles on the big bang. It is solved by the Biblical descriptions here.

Light has another interesting property but I will start with sound to give a basis for understanding. Sound has a property called the doppler effect. Most people have noticed it as a car or train horn goes up in pitch as it comes toward you. What happens with each successive sound release is the train or car is coming closer. As the train passes, each sound release is further away from you. Where there is no air to transport the sound disturbance, there would be no sound. This does not happen with light as it is an electromagnetic wave moving much more rapidly. So colors on trains or cars do not change as they pass. Where the colors would change is at high relativistic velocity or in a gravitational field. If a star or galaxy is moving at high velocity- a good fraction of the speed of light towards us, then the star or galaxy will be closer with each light wave release and the light will be blue shifted toward the higher frequency end. There is only one galaxy coming at us in the Universe and that is our neighbor Andromeda. The rest of the galaxies are moving away from us and so their light is red shifted in the basic elements they possess like hydrogen and helium. What was found in the early 1920's was that the further each galaxy (they were called nebulae in those days) was away from us, the more red shifted were the lines from the elements they carried. This was called the Hubble law as Edmund Hubble in 1929 identified those fuzzy objects called nebulae as galaxies.

The same is true in gravitational fields. Light climbing out of a strong gravitational field will lose energy and have frequency being slowed (longer waves) toward the red end of the spectrum. Light waves falling into the gravitational field of neutron stars and black holes will gain enormous energy and frequency toward the blue end of the spectrum. They will probably end in the gamma ray (γ $-$ ray) end which are the most energetic rays known.

The question always arises. If one rocket can't exceed the speed of light, may two rockets moving toward each other will exceed the speed of light? We will start with each rocket ship moving toward each other at eighty percent the speed of light '0.8c". Using Galileo's addition one might think that the rockets approach to each other would be '1.6c'. If one remembers the contraction of length formula earlier in the chapter then we have a different problem as two rockets are coming at each other with length contraction: [2×0.8c/*square root*(1+4×4/5×5)c×c] which comes out to be 40/41 times the speed of light. You must remember that there is a spatial contraction factor to correct the velocities. If we let β be the fraction of light speed then something is traveling then γ = 1/*square root*(1 − β^2) is known as the Lorentz factor that corrects space and time for relativistic speeds. Only one dimension needs fixing in the direction of relativistic travel as does time. So momentum also needs the same correction. The correction looks similar. relativistic momentum p = mv/*square root*(1 − v^2/c^2). Again this correction is only in the direction of movement. When mass turns into energy it is the old Einstein formula $E = Mc^2$.

The person who contributed most to the formation of special relativity besides Albert Einstein was Hendrik A. Lorentz. Born in 1853, he was a gifted pupil early on. He entered the University of Leiden in 1870 and finished a BS in 1871. In 1875 he finished his doctoral degree on the reflection and refraction of light. In 1878 he was appointed the Chair of Theoretical Physics at Leiden. He clarified the connection between light and electromagnetism. He developed the equations for length contraction at relativistic speeds above. His Lorentz transformations allowed work on the speed of light with distances and time in a single direction. He also posited a theory how electricity flows in wires along electrons. He believed in an ether from which light travels in outer space.

CHAPTER 5
QUANTUM MECHANICS

A basic property of all objects is its ability to emit and absorb radiation. This process is called thermal radiation as it involves an interchange of radiation energy in the electromagnetic field around each object and the thermal motion energy of the object particles. When objects appear with certain colors this means that they are reflecting this color and are absorbing all the other colors in the white light shined upon them. An incandescent solid glows red hot when heated. When heated further it glows white hot which shows some function of temperature. Actually the object emit and absorb at all frequencies and a particular range of temperatures tend to prevail.

Quantum theory began in 1900 with Max Planck. However prior to that in 1859 Kirchoff had showed the total photon energy emitted is divided by frequencies. Therefore the spectral radiant emission per unit frequency per area and per time divided by absorption at that frequency should be a constant based on photon frequency and temperature. The Stefan-Boltzmann law stated that the light energy emission from a hot glowing object was a constant times temperature to the fourth power. A perfect absorber of light does not reflect any light shone on it and is called a black body. Black body radiation was explored many times until the turn of the century at many different temperatures. The radiation energy emitted from a black body has been shown to have a peak at a certain temperature and frequency then reduction in energy output despite increases in frequency and input energy Experiments showed various temperatures and frequencies led to higher peaks with temperature or frequency. With regular heated objects, light emission did not increase beyond the blue-white light range. This was known as the ultraviolet catastrophe. In 1894 the physicist Wien came up with a formula allowing one to calculate the light emission at the peak if one knew the temperature.

At the time there was no known reason why the equation worked but it did except at low ultra-violet frequencies of light.

Black body radiation can also be generated by electromagnetic waves bouncing around in a cavity. When the cavity has a small hole allowing the radiation to escape, we can observe this radiation still with the same peak or Planck spectrum. An analog which is helpful is a string which is tied at both ends. Only certain wavelengths are possible, which are even fractions of the length between the knots. The same is true for black body light waves. Scientists presume that the black body radiation of the big bang was made by a fireball. An alternative explanation was that it was made in a cavity. It would give the same pattern as a hot core inside a cold dark matter shell like Biblical Genesis. Rayleigh-Jeans made a law essentially stating that the power (energy per time) was constant for every frequency interval of light. As frequency of light increases so should the power or intensity.

The first experiments related to quantum mechanics actually were done by Thomas Young in 1831. He performed a double slit diffraction experiment which gave an interference pattern. Whether it was done with photons or electrons one at a time, the question remains how one particle at a time can interfere with itself to give an interference pattern with fringes of light and dark. To start this requires how a radiation field is defined. Putting a boundary shows that the electron stops being a particle and acts like a wave. One has to realize that particles are not like ball bearings. How the slit experiment is done with a single photon or electron at a time is difficult to understand. The photon is a bunch of electromagnetic waves but how can an electron be in two places at the same time to interfere with itself and cause a fringed pattern. What is needed is a better understanding how to make an electron or a photon. How can a quantum object be capable of carrying information about its own structure? My friend John Wallace has written a book about Quantum Mechanics. He agrees with Albert Einstein that quantum mechanics as formulated is incomplete. If one looked at the film where the electron or photon strikes, there would be a small atomic region where the energy is deposited that is not at all fuzzy. This gives the impression that one is dealing with a composite particle.

One would think that electricity and magnetism are completely understood. In 1831 Michael Faraday put together a transformer using a six inch diameter ring of iron as a transformer core. This produced an electromagnetic impulse. Maxwell's description of electromagnetic waves was made in open space. When well annealed iron is used as a coupling agent, there are large scale quantum effects. There are the dynamics of ferromagnetic permeability to electromagnetic waves which is time and temperature dependent and related to the annealing of steel. The dynamics of a magnetic field on such steels is not explainable even today by quantum mechanics. Problems in the dynamic ferromagnetism originate in quantum mechanics and the way relativity was integrated into it.

Early in 1900 Max Planck was forced to explain why the there is a peak in light intensity in black body radiation. By October of that year he reasoned that the entropy of the radiation had to depend on the frequency of radiation where more frequency meant more energy. He also saw energy decreased with the decrease in frequency. He tried to combine the two relations in one expression. The result was an equation relating the frequency of radiation to the energy of radiation. To match these peaks, he was forced to postulate that electromagnetic radiation could only be absorbed or emitted in discrete packets of energy called quanta. The energy of each quanta was a constant times the frequency of the light. This constant became known as Planck's constant and given the letter 'h'. These quanta were later known as photons. The 'oscillators' comprising the black body could absorb energy and could emit radiation only in discrete amounts. They had to statistically distribute these quanta each containing an amount of energy proportional to its frequency $E = hv$ where v is the frequency. This was true for all the 'oscillators in the black body. Planck had to give up his idea that entropy of the second law of thermodynamics was an absolute term but actually depended on statistics as Ludwig Boltzmann had postulated. Thus the world became statistical and of probabilities. Due to the existence of a non-zero Planck's constant, the micro physical world could not be described by ordinary classical mechanics. A revolution was begun by a reluctant revolutionary.

The photoelectric effect was actually discovered in 1887 by the physicist Heinrich Hertz. Hertz found that shining an ultraviolet light on two metal conductors, already with a voltage applied across them, changed the voltage. In 1902 Philipp Leonard found that focusing light on a metal freed the electrons. Work on this phenomena showed that the electron release was not related to the intensity of the light shone on the metal, which would have been the classical explanation. There was no change in electric potential as the light energy increased. The light intensity did increase the number of electrons as measured by the current. Another puzzling part was that the electron flow seemed to start the instant light was shone without any delay. In 1905 Albert Einstein formulated the corpuscular theory of light. This explained that each particle of light, later named photons, contained a fixed amount of energy or quantum. This energy was proportional to the light frequency in waves per second. The proportionality constant was Planck's constant 'h'. Einstein correctly assumed each photon of light would transfer its energy to an electron. With the added energy, the electron would be able to escape the metal with some loss of energy he called the work function. This idea was radical and not fully accepted until 1916 when research by Robert Millikan showed that Einstein was correct.

In 1922 Arthur Compton measured increases in the wave length of X rays after they struck electrons. The energies X rays lost with decrease of their frequency was found to increase the energy of the electrons. Thus it was shown that X rays, discovered in 1895 by Wilhelm Roentgen were made of high energy photons and were electromagnetic waves like light.

For many years scientists have been trying to unify quantum physics and general relativity. In the late 1950's Richard Feynman had proposed a thought experiment to produce quantum gravity and explain the big bang. General Relativity produces a specific strength for the gravitational field of each mass. In quantum theory, particles can not have exact positions and velocities at the same time. A particle can exist in a superposition of multiple places and states and you get only a probability where it will be. Until it is observed, there will be only probabilities. Feynman imagined a mass in two locations. The mass will fall in a gravitational field according to its strength. Thus the gravitational field must have two configurations at once in this interaction and be the hypothesized quantum gravity. If so, possibly quantum theory prevails or possibly general relativity prevails and quantum theory only applies at certain lower energy scales. Feynman wanted someone to come up with an experiment which rules out one of the possibilities. Most physicists think they need a theory of quantum gravity to rule out these possibilities and explain the big bang. I show that the Bible explains the big bang including the quantum effects of stopping a gravitational collapse with just highly squeezed neutrons and protons. The spirit of God hovering over the neutron superfluid waters means that gravity is eliminated. Where general relativity is nonexistent is density of $10^{17} grams/cm^3$. This is the core density of the smallest black holes, just over nuclear densities. Particles can not move and there is no gravity despite quantum jitters. This is the interface between quantum gravity and general relativity and where the Universe stopped collapsing and bounced. Quantum gravity here means no gravity.

CHAPTER 6
HISTORY OF GENERAL RELATIVITY

Albert Einstein, as genius as he was, still needed 2 friends who contributed extensively to both Special Relativity and later General Relativity. Einstein met them both while in the Swiss Federal Polytechnic School in Zurich and both were life long friends. The first was Michele Besso. Besso was an imaginative mechanical engineer whose extensive discussions on Special Relativity with Einstein earned him the sole recognition on Einstein's 1905 Special Relativity paper. Special Relativity is concerned in one direction with the expansion of time and contraction of space. In 1907 Einstein began thinking about General Relativity while he still worked at the Swiss Patent office. Einstein was interested in a theory which would account for the precession of the Planet Mercury when it was closest to the sun. It precessed meaning that it came out of close contact with the sun about 43 seconds early every century as a result of space contraction. With a new theory Einstein had to explain a number of other phenomena. Galileo's principle had all matter falling in a gravitational field at the same rate. Einstein appreciated Mach who stated there was no absolute space unlike Newton but rather distant stars effect local matter. Inertial mass is the resistance of matter to acceleration. Gravitational mass determines the force exerted on a body in a gravitational field. These are always equal, although the reason was unclear. Einstein knew from Special Relativity that inertial mass is energy and that the gravitational mass must depend on energy as they were equal. Also gravitational theory must maintain Galileo's principle. Einstein tried to generalize special relativity to four directions of spacetime but that required giving up Galileo's equivalence principle. Einstein took some guidance from the electromagnetic field theory from Lorentz. He showed how space is filled with electric and magnetic charges which form a field. The electric field in space is the forces acting on a unit of electric charge. The matter in an electric field is like a source of the field and the field determines how matter moves. Although this is somewhat similar to a gravitational field, electromagnetism was then assumed to be operating in a special relativistic way in one direction. Einstein had the most difficult time applying this to curved space-time. Fortunately, Einstein never gave up on his equivalence principle, where an acceleration in an elevator is equivalent to gravitational forces.

Einstein only made significant progress in 1912 when he returned to Zurich and was re-united with another old friend Marcel Grossman. Grossman was a mathematician who had learned about Gauss's theory of curved surfaces and Riemannian geometry of curved space. Grossman taught Einstein the mathematics. Their work from mid 1912 to the beginning of 1913 was correct but they didn't initially understand the new spacetime geometry. Einstein and Grossman gave it up even though it explained the precession of Mercury. The notebook which contained the blueprint for a full tensor theory was called the Zurich notebook and Einstein called it 'Relativat'. The full theory contained new terms called Christoffel symbols introduced by Elwin Bruno Christoffel in 1882 and incorporated into Riemannian curvature by Gregorio Ricci-Curbastro and Tullio Levi-Civita in 1901. Initially neither Einstein nor Grossman nor anyone realized the importance of these terms. Although they are not tensors, Chistoffel terms (Γ) measure the bending of spacetime coordinates. They are constructed from 3 metric tensors each measuring local coordinate curvature. The bending of spacetime coordinates is actually the effect of gravity. When these Christoffel terms are zero, there is no gravitation. Then the metric tensors are Euclidean and the coordinates are straight lines. Not understanding this at first, our heroes went back to an earlier theory of Einstein called the Outline or Entwurf theory. The 'Entwurf' theory was incorrect as it also took no cognizance of Christoffel symbols and their significance. In 1913 Einstein and Grossman felt the Entwurf theory was the best they could do. It didn't solve the precession of Mercury. There were other things Einstein felt he had to incorporate into the gravitation theory. The equivalence between gravitation and acceleration was foremost.

The conservation of energy and momentum had to be accounted for. In basic mechanics there are three separate equations to account for them. In special relativity they are combined into a single tensor called the energy-momentum tensor which accounts for changes in one direction. Somehow changes in 3 dimensions with time had to be accommodated. These had to reduce to Newtonian equations at low velocities or weak gravitation. Einstein recognized by 1912 that gravitation is due to spacetime geometry. A particle following only gravity would move in the straightest possible path in spacetime. This path is called a geodesic, when not disturbed by other forces. To calculate this, the left side of the equation is an acceleration (second derivative of position respect to time) of the particle but using the particle's proper time tau (τ). On the right side is the Christoffel coefficient representing the gravitational field by bending of spacetime. It is followed by the rate of change of 2 dimensions. When the Christoffel coefficient is zero, the acceleration vanishes and the particle is at constant velocity or not moving. At slow non relativistic speed, the proper time and surrounding times are the same. Here the equations reduce to Newtonian terms. Einstein stated that recognizing the importance of the Christoffel coefficients were his greatest challenge. It seems no one at that time understood them. Finally by the end of November 1915, Einstein and Grossman finally put everything together correctly. Mathematician David Hilbert had heard some of Einstein's lectures. There was some question whether he might have beaten Einstein and Grossman to Relativity. However in hind sight we know he didn't understand the importance of the Christoffel terms either. His galley proofs from early in November 1915 showed he had missed this critical point. He did approach General Relativity differently. He developed the least action integral from Relativistic processes. An example of this means that free fall particles take the shortest path in spacetime between two points. An airplane traveling some distance over the earth takes a geodesic path which is the shortest path and is a curve. David Hilbert's Least action integral is very useful in many situations involving general relativity. He was a 'mentsch' in several regards. One is he never made any claims about getting to Relativity first. Secondly the Nazis came and chased out all the Jewish scientists. They asked him how the mathematics and physics research was going. He responded to them that there wasn't any(you've gotten rid of the Jewish researchers).

Some additional points on general relativity are important. There is no prior geometry in spacetime. Spacetime will curve near masses. The larger the mass the more curvature or bending it causes. As the mass moves away, the curvature is reduced almost like a stretched rubber sheet. Einstein's happiest thought he recalled during a lecture at the University of Kyoto in 1922. When he worked at the patent office in Bern, he realized that a person falling will not feel his own weight. He later stated this was the happiest moment in his life. He began his thought process that a gravitational field will cause the same acceleration to all bodies. Thus gravitation and acceleration are essentially the same. A person in free fall does not feel gravity. The equivalence principle helped Einstein before he could put a full gravitational theory together. A beam of light shined across an elevator will be deflected downward in any relativistic acceleration as the elevator travels upward. Einstein postulated that a gravitational field also must bend light. In 1919 an eclipse was observed by Arthur Eddington did find stars nearer the sun which made Einstein and general relativity world famous. What happened is the light rays were bent in the sun's gravitational field, curving them around the sun. Thus stars physically behind the sun become visible when then sun's massive luminosity is temporarily eliminated. In classical mechanics, an inertial mass is the resistance a body has to acceleration. Mass also determines the force on a body in a gravitational field. This is called the gravitational mass. It had been shown to great accuracy by the physicist Etovos that the two masses are equal. When Einstein realized that the equivalence principle was the cause of this equivalence, he had his happiest moment. One must realize that gravity causes non-Euclidean geometry. When there is no gravity then the Christoffel coefficients are zero and Euclidean geometry with straight lines is the rule. A rotating mass will cause electromagnetic waves in radio, TV or cell phones and gravitational waves.

It was interesting that astronomers in the '90's tried to determine which geometry we had inside the Universe. They measured what were felt to be standard candles-the supernova star explosions of

Type 1A. If the geometry of the Universe was closed, then the supernovas would seem closer to us than expected. If the Universe was open, then the supernovas would seem further away. My paper at the end of the book explains this technically. What the astronomers found was that this type supernova measured less luminosity and looked even further away than expected even for an open Universe. Instead of just open, what they measured seemed like every place in the Universe was accelerating from every other place. This was called dark energy. It seemed stronger than all the gravity in the Universe. According to this result, our Milky Way Galaxy would be left alone in an ever expanding Universe to die a cold death as the stars run out of energy. Do not worry. The Bible tells us the Universe did not start as a fireball and shows that former black holes like the big bang will run out of gravitational energy. The scientists' imaginary Big Bang fireball means that scientists can not get out of its gravity. There is no scientific solution today for a fireball that started the Universe nor the dark energy. The Biblical black hole that started our Universe lost all its gravity during contraction. This allowed the Universe to start unhindered by gravity. Not only did the initial super massive black hole lose all its gravity, but the black holes or firmament (REKEA in Hebrew) at the center of each galaxy, are all losing gravitational energy as they collapse as time passes. There is no singularity of infinite gravity in each black hole but a central area of zero gravity as the particles are unable to move. So galaxies are further away from each other due to gravitational loses not because the Universe is expanding faster and faster. Once the distant galaxies begin contracting, their light will be blue/violet shifted not red shifted as currently. The Universe will end up in a Big Crunch and start all over all in another world. The Midrash of Jewish literature explains that there were previous Universes but God didn't like them because there were no humans.

The person who contributed most in applying general relativity to the Universe was Aleksandr Aleksandrovich Friedmann. Friedmann was born in St. Petersburg Russia in 1888. His parents divorced when he was 9. He was in St. Petersburg Gymnasium and soon was one of the top pupils along with his friend Yakov Tamarkin. In

1905, the two friends submitted a paper on Bernoulli numbers to Mathematische Annalen with David Hilbert as the editor. The paper was published in 1906. !n 1907, the two boys attended a modern physics seminar run by Paul Ehrenfest. Ehrenfest was a brilliant statistical physicist who unfortunately also suffered depression. Friedmann, Tamarkin and Ehrenfest discussed problems in special relativity, quantum and statistical mechanics. Friedmann's father died prior to his completing his undergraduate studies in 1910. Despite financial hardship, he continued studying for his masters degree. He lectured on elasticity, hydrodynamics and later meteorology. After the first World War started, Friedmann volunteered to fly aircraft. He was soon involved in dropping bombs and modeling their trajectories. While in Kiev, he gave lectures on aeronautics for pilots. When the Central Aeronautics group moved to Moscow he went there in 1917. When the Communists took over October 1917, they closed the Aeronautics Station and group down. In 1918 Friedmann was selected to be an extraordinary Professor of Physics and Mathematics in University of Perm. There Friedmann started the Institute of Mechanics and was elected to the Editorial board on the new journal Physicomathematic Society. The city soon was caught in the struggle between the White Russian Army and the Red Russian Army. In early 1920 Friedmann returned to St. Petersburg now called Petrograd by the Communists. He began teaching mechanics and mathematics at Petrograd University. He also taught physics and mathematics at the Petrograd Polytechnic University. In late 1920s, Friedmann wrote a letter to Paul Ehrenfest that he was working on relativity using two principles. First involved uniform motion in normal geodesics. The second involved the constant velocity of light outside of the one dimension of Lorentz transformations. In June 1922 Friedmann wrote a paper on the curvature of spacetime in the Journal Zeitschrift fur Physik. He noted that stationary Universes were the ones considered by Einstein and de Sitter. He showed how general relativity also supports cases where the radius of the Universe is increasing/decreasing or changing. In September 1922, Einstein responded to the idea of a variable Universe with disapproval. He stated that Friedmann's work was suspicious that it does not satisfy the field equations. By December 1922, Friedmann wrote back with his now famous Friedmann equations. Einstein recanted and these are now widely accepted as the basis for relativity applied to the Universe. Although they are described with their limitations in my technical work at the end of this book, I will attempt to describe them here non-technically. Friedmann had two equations for the rate of change of the Universe scale factor (loosely the Universe radius). The first was an acceleration equation of the scale factor with matter. The second equation was a velocity or rate of change of the scale factor with matter and radiation. The first equation can be derived from the second by differentiating it. So just the second equation needs be used. It can contain a term Einstein added to make a steady state Universe. He discarded it when finally he accepting an expanding Universe after viewing Edwin Hubble's work on galaxies. Friedmann's equations are still used today for calculations in the Universe. The problem I have with the Friedmann equations is that they assume a nonexistent perfect fluid, one that stays uniform despite changes in pressure, density, etc. Scientists think general relativity breaks down at the start of the big bang but is doesn't work for black holes either.

CHAPTER 7
COSMIC BACKGROUND RADIATION

The cosmic background radiation was discovered in 1964 by two Bell Labs scientists Penzias and Wilson in Homdel New Jersey. They had built a large radio receiver for another project. Once it was completed, it kept getting a background radio noise. They tried cleaning the pigeon dung and other things out of it but could not get a clean signal. The prior year in Princeton New Jersey a Professor of physics, Robert Dicke had instructed his two graduate students Peebles and Wilkenson to build a receiver to find if there was evidence of a big bang in radio waves in the Universe. It was known that the Universe was expanding since the days of Hubble 30 odd years prior. Hubble had found that the further away galaxies were, the faster and light more red shifted they seemed to going away from us. Light waves would lose energy as the Universe expanded and a big bang flash would now be in the lower energy radio wave range, if it existed at all. Once the Bell Labs guys made a phone call to Princeton about finding this long wave radio noise, there was great excitement. Two papers were published in the 1965 journal Nature. The first was the Bell Labs scientists describing their receiver which lcd to their Nobel prizes. The second paper described the radio waves in a Planck spectrum as the evidence for a big bang and the end of the steady state theory of the Universe. Fred Hoyle was an astronomer and major proponent of the steady state theory of the Universe before World War II. He had derisively coined the name Big Bang for the start of the Universe many years prior. Suddenly there was proof for it. Starting with Snyder and Oppenheimers' paper in 1939 that continued gravitational collapse would lead to singularities and nearly infinite energies, virtually all scientists were convinced that a fireball was the start of the Universe. Snyder and Oppenheimer had started general relativity equations for matter with plenty of room to move at reasonable densities. They extrapolated gravitation equations mathematically (this assumed rapid movement of smaller and smaller sized nonexistent particles) until the infinite density of a singularity. This was widely accepted despite some scientists like Eddington having reservations about getting to a singularity. There was no known force to stop gravitational collapses from reaching very high densities as the physical size of particles was not considered. Neither Infinite pressure nor Fermi energies could stop large gravitational masses (over 4 solar masses) from reaching infinite density. But could gravitational collapses even come close to these densities? With the Planck spectrum found in the Cosmic Background radiation, the belief in a fireball to start the Universe was widely accepted. As a background note, our sun puts out a Planck Spectrum from a surface temperature of 5762 degrees Kelvin. It is close to a fireball. In the 1990's the Cosmic Background Explorer satellite explorer began measuring deviation of the photon temperature from the 2.7 degrees Kelvin temperature Planck spectrum. The work done by George Smoot and his team of scientists showed that the deviations in the Cosmic Background Radiation (CBR) were at a 10^{-5} or one part in one hundred thousand. As the Universe had expanded from the big bang, the last temperature in the Universe that had the ability to interact matter with photons was a little over 300,000 years after the big bang. We are now about 13.7 billion years after the big bang. If there was just unstructured matter in the Universe from a fireball at that time, this would correspond to under ten thousand solar masses. The CBR showed that the start of the Universe had flat spacetime-that the expansion energy exactly matched the gravitation. This was not explainable by the singularity and fireball picture the scientists had. It looked more like a bounce where some matter stopped the prior collapse and then released that energy into an expansion. There couldn't be any gravitation at the start of the big bang or it would have reduced the expansion energy. To get out of the enormous gravitation a start around a singularity, in 1980 Alan Guth followed by others invented a scalar which inflated the Universe from near a singularity to much larger size. The problem was that there was no reason to start or stop such an inflation. Many scientists searched the CBR and Universe for any evidence of inflation and none has ever been found. There were subsequent negative searches of the CBR by succeeding satellites for further changes or polarizations due to inflation. Another problem is that no connection of the Cosmic Background Radiation to present day galaxies or even galactic clusters has been found. In addition it should be noted that the Biblical model of a hot core surrounded by a cold shell makes perfect Planck spectrum radiation as well. The standing waves of photons between a hot core source and the cold shell absorbing produce a fine Planck spectrum.

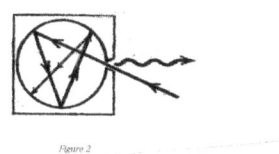

Figure 2

Figure 2: **Making A Planck Spectrum**

The Bible describes a cold shell and hot core. Scientists believe in a fireball which can make radiation with a Planck spectrum. Also standing radiation with a hot core and cold shell can make a Planck spectrum which is not related to galaxies in the Universe. The background radiation we have today is not related to any of the matter in the Universe.

CHAPTER 8
WHY GENERAL RELATIVITY FAILS AT HIGH DENSITIES

It has long been known that general relativity fails to work well in several areas of the Universe. Neutron stars are the lowest density where things don't work well. It is well known that it fails in the big bang going back to a supposed small fraction of a second after the big bang. It is not realized that it also fails in black holes because there is no way to directly ascertain whether the core of the black hole has a singularity or not. This is true despite the big bang and black holes supposedly both having almost infinite matter densities.

In order to cause a curvature of spacetime which causes gravity, there must be matter and energy. All matter is related to energy by Einstein's famous equation $E = mc^2$. The energy of matter is measured in both its movement and its quantum waves. As detailed exactly in my paper at the end of the book, the stress-energy tensor contains the linear velocity of particles, its rotation and curved motion, thermal jiggles, vibration and quantum spin and quantum jiggles. With the possible exception of quantum spin and quantum jiggles, all movement causes spacetime bending or gravitation. Much like an isolated light bulb sending out light waves in all directed, matter and energy will cause gravity in all directions. Gravity expands in all direction. If we imagine the surface of a sphere, gravity will lessen as the sphere expands. Since the sphere surface is related to $4\pi r^2$, for each radius r, the strength lessons by $r \times r$. Doubling the distance from the source will make gravity one quarter strength and tripling the distance from the matter will leave one ninth. It is very important to realize that gravity depends on the motion of particles. While it is true even at absolute zero particles still have some quantum motion called wiggle, highly squeezing them will be enough to cease spacetime gravity changes. This means there must be room for the particles to move. Every school child learns that matter occupies space. It is true most of particles are space. The atom is 99.9% space. The neutrons and protons in its nucleus also contain 99.9% space before you get to the tiny quarks in each. We do not believe quarks are composite particles and it is extremely difficult to pull them apart due to the strong nuclear force. Amazingly the strong force gets even stronger as one pulls quarks apart. Cosmologists extrapolate general relativity by using just using mass density and pressure for gravity all the way to singularities. To get to this mass density, there is a presumption that particles are moving enough to cause spacetime changes, regardless of particle size. Pressure adds to gravity and also pressures presume the particles are moving.

So general relativity works perfectly at densities below levels of plain neutrons and neutron stars about a hundred trillion (10^{14}) grams per cubic centimeter. With all the matter in the Universe in one place to start, they know General Relativity fails. They don't know why and just continue extrapolating mass density and pressure. General relativity fails because they can't pack any more matter than about one thousand times that of a neutron star into one place. This is the limit black holes have and it limits them to about 4 solar masses in size. If more density could be packed into one place to bend spacetime around, there would be smaller black holes. Sometimes they employ negative pressure to try to get out of their extreme densities and gravity for the big bang. On our level negative pressure makes no sense. Neither does using mass density and pressure to the infinite densities of black hole singularities. I put papers in showing that extrapolating general relativity like that doesn't make sense and violates the rules for construction of the stress-energy tensor. My papers were rejected summarily or with the comment that I can't challenge general relativity. I tried rephrasing the construction of a stress tensor using the accepted authority Meisner, Thorne and Wheeler's book on gravitation. The paper was still rejected. Is there proof that astrophysics objects like neutron stars have lost gravity? Not only neutron stars lose gravity-you can check my scientific paper at the end. My paper also includes evidence that black holes have lost and are losing gravitation. Heresy! You wouldn't expect that if black holes have a real infinite gravity singularity inside them.

Scientists are so in love with their 'infinite' gravitational power, they don't want to hear about evidence contradicting that. I had posted a number of papers describing a much lower power big bang on the internet astrophysics archives in the late 1990's and early 2000's. They garnered only some nasty comments or were ignored. As a result of Jefferson Labs showing in 2018, that protons are extremely difficult to break into quarks, it is obvious then that these particles can't move at high densities. Even if the neutrons and protons broke into their up, down and possibly other quarks components, at one thousand times more in density. these quarks would not be able to move either. So there is no such thing in the black hole as a singularity but rather a core of zero gravity from highly squeezed particles. As time goes on, the zero gravity core will expand until there is no gravity for the entire matter. It will then break up and return all black hole matter to the outside universe as quantum theory demands. It is similar to the big bang although black holes radiate much of their energy away during formation. It takes at least four times the mass of the sun to produce black holes. No lesser masses have the gravitational energy to twist spacetime back on itself like a black hole. This is the limiting density in the Universe. Smaller black holes which require higher densities to form can not exist. Steven Hawking had claimed that small black holes, much less than the size of the sun would emit radiation at their Schwarzschild radius. This radius is the boundary where half of a particle-antiparticle pair may get trapped and the other would lead to radiation. This was called Hawking radiation. It has never been found as there are no small black holes under four solar masses. Dark energy is due to loss of gravitation in galactic black holes, nothing more. The galaxies are further away than they should be compared to the expansion of the Universe. Since all the matter in the big bang must have been a black hole, we wouldn't be here if it didn't lose gravity at high density.

CHAPTER 9
WHY THE BIG BANG MUST HAVE ORIGINATED IN A
BLACK HOLE

As I mentioned above, I never believed in singularities even before I studied general relativity over 35 years ago. The scientific idea of a singularity really started in September 1939 with J. Robert Oppenheimer and H. Snyders' paper 'On Continued Gravitational Contraction'. They applied general relativity to a collapsing mass and presuming the stress-energy tensor was unchanging in pressure and mass density. They mathematically extrapolated general relativity to infinite density and gravity. No one knew any reason to stop a gravitational collapse before a singularity either until recently. No one before myself has ever really challenged the idea of an infinite collapse of sufficient matter. The problem was that there was no evidence of such an infinite collapse either. The end result was a black hole which is a hidden result. It is not that gravity is so strong that light can not escape. Rather a black hole means space is bent back on itself like a hair pin as time is dilated. Since space and time have opposite signs it must be such. There is no current possibility to examine the results. There is no intermediate steps prior to a black hole to examine incomplete collapses. Still I was unconvinced from before I talked to Professor P.J.E. Peebles at Princeton in 1991. I felt that the big bang must have been a black hole due to all the matter in the Universe being in one place. I formulated my model according to the Bible not reaching the tremendous energies or densities that would cause a fireball. At the time in 1998 I didn't have an explanation how to stop a gravitational collapse at squeezed neutron and proton densities but knew it fit a pattern for galaxy formation with normal matter as cold dark matter, etc. My specialty and history in engineering modeling and current job in biological pattern recognition allowed me to put together a dark energy solution even prior to the problem being found in 1998 by two groups of astronomers 'Reiss and Perlmutter'. What they were doing is trying to determine the geometry of the Universe from our inside location. They used a type of supernova 1A as a standard candle. If there was more than enough matter causing gravity to close the Universe, then space would be bent inward and the supernovas would seem closer and brighter. If there was just enough matter currently to close the Universe, the supernova would just the right luminosity expected. If there was less luminosity, the supernova would be fainter but within an amount depending on how short the matter was to close the Universe. What both teams found was that the supernova luminosity was even further away than no matter at all in the Universe. After rechecking their results multiple times, they announced that some strange energy was pushing the Universe apart. It was stronger than all the matter causing gravity. Although no one would listen to me, I had placed on the Internet Astrophysics Archives in 1998 and 1999 the solution. Black holes in the center of the galaxies are losing gravitational energy. To people who were taught and believed in singularities, I was like a nonbeliever in the middle of the Catholic Church. My papers received nasty comments or were ignored. Even Rocky Kolb, who I worked with those years on nucleosynthesis, was politely dismissive. Although I wrote 6 total papers those years on the internet (see below), they were mostly ignored. I even titled one about normal matter being the (not found) dark matter with the same results. It wasn't until 2011 I found papers on neutron stars losing gravitation and added them to my scientific paper. A few years later there were articles about matter headed right for black holes and then later traveling away and even black holes having only a tiny fraction of the magnetism these should have. Adding these to my paper did not help it get accepted anywhere. It was now more than 20 years later and there still is no accepted solution to dark energy or dark matter. In 2018 was a study of protons by Jefferson Labs in Newport News, not far from me in Norfolk Virginia. J-Labs found that protons have enormous resistive pressures against collapse. A normal proton may be 0.8-1 femtometer size (a femtometer is 10^{-15} of a meter or one over a 1 followed by 15 zeros). If it is squeezed to 0.3 femtometer size, it will take 10^{35} pascals or about 10^{30} atmospheres to crush it. That is a 1 followed by 30 zeros. J-Labs noted in their article that this was more pressure than a neutron star could possess. While it is true that matter collapsing to a black hole could crush the proton to quarks, it would require the matter to be traveling near the speed of light. A regular collapse isn't fast enough and won't do. I maintain the center of collapsing matter is composed on neutrons and protons all highly squeezed and therefore not generating gravity except matter outside the highly squeezed core. One can not crush the protons nor nucleons into quarks. Once densities are reached with particles highly squeezed, they won't be moving at all. That includes no velocity and also no vibration or rotation. There is no data to presume quantum spin is suppressed or not at enormous densities. Is there gravitational loss to support my ideas? As mentioned, there is unexplained evidence of gravitational losses in neutron stars and black holes. I extrapolated this to the big bang which would allow matter to escape without gravity at its start. This would allow the collapsing energy of a prior Universe to squeeze the protons and neutrons but not change them into quarks. As the particles were highly squeezed and unable to move, gravity would disappear and any disturbance could start another big bang. The scientists are currently stuck in a black hole. The evidence for their scalar which inflated the Universe from microscopic is nonexistent. Their fireball which started the Universe doesn't solve much anyway. A fireball would have extreme difficulty matching the expansion energy with gravity. The expansion of the Universe is so finely tuned that it needs a missile from earth hitting a target 3 feet wide on the other side of the Universe 13.8 billion light years away. The galaxies are highly correlated light emission, rotation speed, rotation direction and many more parameters. A random process fireball can not hope to make the highly correlated galaxies. Sure enough the scientists are stumped on galactic origins. See section on galaxy formation for further information on this. All the high energy particles that would remain from a fireball are not found in our Universe such as monopoles, domain walls or even antimatter.

CHAPTER 10
WHAT IS COLD DARK MATTER

Cold dark matter really got a start in 1933. Edwin Hubble had identified the 'nebulae' in distant space as galaxies in 1929. Fritz Zwicky was studying them while working at the California Institute of Technology. He applied mathematics to the Coma Cluster of galaxies. The effect of gravity on the galaxies did not account for the orbital speeds. He realized there was much more matter than could be seen. His calculations estimated dark matter at 400 times the visible matter. He coined the name in English of Dark Matter. This unseen matter was needed to hold the cluster together. Although his estimate of dark matter being 400 times the amount of luminous matter was way too high, he pioneered the study of dark matter. In 1939 Horace Babcock had been studying our closed neighbor galaxy Andromeda. He found that light and presumably mass to luminosity ratio increased from the center of the galaxy outward. Although the reasons he attributed to this effect were wrong, he began the study of dark matter effects on individual galaxies. Beginning in the 1960s, Ford and Rubin were measuring galactic rotation curves. Rubin used a spectrograph which measured the velocity of the outer edge of spiral galaxies looking at them edge on. Rubin found evidence for about six times as much dark matter as there was visible matter. Radio astronomers using Jodrell Bank and Green Bank had found the outskirts of Andromeda, our nearest galaxy, was not reduced in rotation velocity. Kepler's law would mandate slower rotations at greater distance from the center. Later other galaxies were found with same phenomena-flat rotation velocities throughout the galaxy radius with newer equipment. All these measurements suggested increasingly large mass to visible light ratios in the outskirts of galaxies. Dark matter scales as density is proportional to volume. Dark matter has gravitation but is cold and produces no light photons. Non reacting matter is also required for nucleosynthesis of light elements to give correct yields of helium-4 to hydrogen and appropriate yields of their isotopes. Yet there are scant further evidence to what dark matter is.

Our earth is a composite of multiple things: Asteroids, meteorites, comets, inter planetary debris and planetary matter from planets like Mars. I never could understand why we could not find any of this strange dark matter in our solar system. There are other 'particles' that should be in our solar system if the Universe collapsed to smaller than the period on the end of this sentence. These include monopoles, domain walls, antimatter and other things that haven't been found to exist or exist in any significant quantity for antimatter. Antimatter will result from not very extreme temperatures applied to free matter. But if matter and antimatter come together such as anti-electrons or anti-protons, with regular matter this will cause a large energy release as the particles annihilate each other. Assuming free protons and the neutrons there has been a major mystery why there are not even amounts of antimatter in the Universe. In the standard model of things in the Universe, there are electromagnetic fields, scalar fields which have a single value throughout. There are gravitational fields which we live in and depend on to hold our things, houses and ourselves from flying away. There are electrons and larger muon and tau+ particles in the same (lepton) family. There are 3 families of matter. Corresponding to these are electron, muon and tau neutrinos. These very tiny neutral particles which travel almost at the speed of light, interchange with each other in flight and can travel through the earth without an interaction most of the time. There was a puzzle at one time why the sun produced on 1/3 on the electron neutrinos thought to be made there as hydrogen is fused to helium with photons and neutrinos. The puzzle was solved when all 3 neutrinos were found transforming into each other during flight. Our neutrons and protons are called baryons because they have 3 quarks in them. The proton has two up quarks and a down quark. The slightly heavier neutron has two down quarks and an up quark. Of late, these particles may also contain a strange quark and anti-strange quark in them. Quarks and antiquarks do not destroy each other the way larger particles do. As a matter of fact, the strong nuclear force or just strong force holds them together. There is also a weak force which causes radioactivity and can break down most quark combinations except for protons. This force causes radioactivity. Also in the quarks there is charm, top and bottom for a total of six quarks making all. Of the forces there are basically strong, weak and electromagnetism. The Higgs Boson is a force responsible for the matter in the Universe. The large Hadron Collider (LHC) has been searching for additional matter outside what I've listed for you. Many years with more and more powerful energies the LHC has searched for additional types of matter with no success. It did find the Higgs Boson part of the Standard Model in 2012, causing Peter Higgs and 2 associates to get the Noble prize. This is a particle that controls the size of other particles and is known as 'The God Particle'. Its small size of 125 GeV means that there are no unusual other matter created. This fact hasn't stopped physicist from searching. The search has gone until 13 Trillion electron volts (TeV) without success finding any strange matter or unusual matter at all. The proton and neutron can be found at energies around 1 Billion electron volts (GeV). Only standard model particle matter and its antimatter have been found. Since at least 1985 there have been hundreds of searches to determine what dark matter could be. The searches included high in the sky, on and under the earth. In the last 50 or so years, there have been literally hundreds of searches for particles as small as neutrinos which have almost no mass to 100 times the mass of the sun. Anything scientists can think of has been searched for after some fund has been found to pay for it. No one dares question their conclusions or if they did, no reputable journal would publish it. Even the internet astrophysics archives where I have five papers from 20 years ago has moderators which have been intolerant of late in diverse opinion. So I guess this will go on until someone either takes the first part of the Bible seriously, reads my papers with this book or the world gets tired of funding searches and machines that produce nothing. It certainly has been a big industry and many full time jobs for scientists searching. All this searching by most of the scientific world is based on extrapolation of general relativity to Planck distance much small than an atom and expecting it to be correct. They want to spend trillions of dollars to build a new particle collider much more powerful than the current Large Hadron

Collider (LHC) now on the French-Swiss border. The current one has not found any unexpected new particles except the Higgs Boson in 2012 which was expected. This is why the physics magazines could never accept my papers. It would put hordes of theoretic and experiment physicists out of work and ruin their industry. All the big geniuses can not change Biblical reality so they ignore it.

CHAPTER 11
WHAT IS NUCLEOSYNTHESIS

N ucleosynthesis is one of the few bright spots in science's understanding of the Big Bang. Whether the Big Bang either started at almost infinite temperature with quark soup and cooled as it expanded or started with neutrons or protons the synthesis of light elements of hydrogen, helium and their isotopes is a success story. Once the Universe cooled to around 10^{10} (10 billion) degrees, neutrons stopped being formed. Neutrons by themselves will break down in under 15 minutes to protons, electrons and a neutrino. So neutron numbers declined compared to protons from 1:1. As things cooled to about 1 billion degrees there were only 164 neutrons for every 1000 protons. At this lower temperature neutrons began to combine with protons. Deuterium or heavy hydrogen is formed by 1 proton, 1 neutron. Neutrons do not break down once in combinations. Deuterium is very reactive in this temperature range. It was not made later in stars but destroyed. Also controlling the reaction rates to deuterium and heavier elements was the photon number. There were about 1.6×10^9 (1.6 trillion) photons of light for every baryon (neutron or proton). Here things were kept reacting by a large amount of nonreacting matter $\approx 95\%$ (Science: dark matter or Bible: cold normal matter) slowing the Universe expansion. From deuterium was produced Helium-3 with 2 protons and 1 neutron or Tritium with 1 proton and 2 neutrons. From these two elements came Helium-4 with 2 protons and 2 neutrons. Helium-4 is quite stable as it has the highest binding energy of all these light elements. So as things cooled during further expansion, Helium-4 comprised about 25% of matter from the big bang. Only tiny amounts of Deuterium, Helium-3 and Tritium remained. Even tinier amounts of heavier elements such as Lithium-7 were produced. Stars actually have made everything else. They fuse hydrogen into Helium and heavier elements. This process produces energy and sun light until iron. Elements heavier than iron cost energy to produce which is why there is much less of them. Fred Hoyle, the astronomer who derisively named the Big Bang, figured out how Carbon-12 was made in stars from three Helium-4 atoms. The 'God' particle, a Higgs Boson is supposed to give other particles mass but is too small (125 GeV) to make any of the high energy non-existant particles that the European large Hadron collider (LHC) has been searching for.

The presumption has been that normal matter would have reacted due to heat during nucleosynthesis. This would produce helium, gravy hydrogen and other elements. If normal matter wasn't very hot, it would be the dark non reacting matter that scientists can't find with their nuclear colliders or dark matter searches. This would produce Helium, Deuterium and other light elements.

CHAPTER 12
ABOUT GALAXIES THAT FIRE BALLS CAN'T MAKE

O nce scientists start with a fireball for a big bang, highly structured galaxies are most difficult to make. They start with small masses (over densities which form by gravitational collapse), much less than the size of the sun in some mathematical distribution. In order to grow the masses, there must be mergers, gas accumulated on small black holes, gas being expelled by black holes and some mechanism from adjacent structures to control matter accumulation. One must realize that these are all random processes. Galaxies do congregate in clusters, along long lines called filaments and are absent for large volumes of space called voids. On the cluster and supercluster level, galaxies do have a somewhat random spacing. However galaxies themselves are highly correlated on circular velocity, light emission, direction of rotation, etc.

The current scientific mechanism of galaxy formation starts with a dark matter halo forming over random small masses. This over-density acquires more matter through gas accumulated on small black holes or mergers of the black holes accumulating additional dark matter orbiting. The gas contracts and begins rotating. Inside the newly rotating matter, between the forming stars, empty spaces are involved with cooling and a feedback process controlling the size of the forming stars. This process also causes ejection of some of the dark matter halo through winds. Gas flowing to the center of the galaxy will be captured by the central black hole in a process called accretion. Other interstellar material will be transferred through winds or the jets of accretion. The density of the halo dark matter is affected by these processes and well as local unattached dark matter. Once the gases have cooled, collapse to higher mass stars can begin. Super massive central black holes can grow from matter expelled from small black holes, other matter not gravitationally attached to large masses and mergers. Galaxies growth is mostly dependent from

58

feedback of stars and black holes. Ionizing radiation from young stars, supernova star explosions as well as black holes accreting mass and releasing much of its energy will cause much galaxy mass and energy outflow. A problem arises in that too much dark matter is left at the center of dwarf galaxies where there is no evidence for dark matter. Larger galaxies with rotation rates over 100km/sec have dark matter evenly spaced from the center outwards. Thus there is less dark matter than expected on simulations. Dark matter and

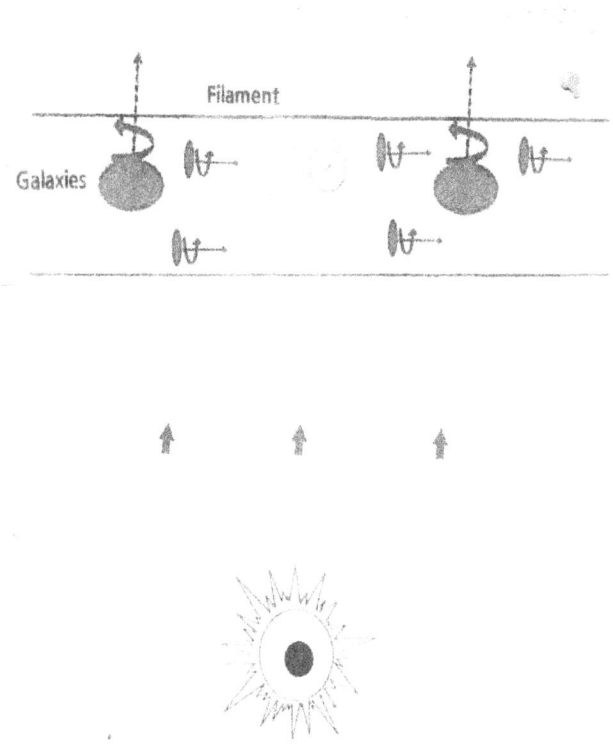

Figure 3: **Biblical Galaxy Formation**. Large and small shell masses were driven into the universe forming filaments. Larger masses coalesced into black holes which held a fairly even density of smaller dark matter. Later hot core gas was captured according to the depth of the gravitational well. During capturing process, larger black holes, unlike smaller black holes, did not change the direction of rotation.

normal matter follow similar unexplained profiles which must be highly adjusted to give an unchanged velocity from the center of the galaxy to its outer visible areas. The high galactic angular velocity is unexplained starting from hot gases and small masses. The dark matter galactic halos are supposed to grow by attracting surrounding matter. The central black holes are supposed to grow by accreting matter from stars that die and become exploding stars called supernovas. However this process would make much larger central black holes than are present. Galaxies are supposed to grow by mergers yet many of the large disk galaxies show no evidence of mergers. The intergalactic spaces of the Perseus Cluster of galaxies show a uniform distribution of iron. This must have occurred prior to the formation of the cluster and its supernovas.

From a side view in the galaxies that have central bulges, the mass of this bulge correlates with the size of the central black hole and average velocities of the bulge stars. It is unknown why the gas that formed the bulge stars settled near the black hole. Why does the inward bound matter go to the bulge in some galaxies and to the central black hole in others? Accepted theory has galaxies with and without bulges accreting matter during the period that massive early stars were forming. Galactic bulges do contain old stars which contain many heavier elements than Hydrogen and Helium. There seems no reason why old stars avoided galaxies without bulges. The old stars do not appear in the dark matter halos and must not have played a role in galaxy formation. The mass distribution in spiral galaxies is evenly spread from the dark matter in the halo outer limits to the inner baryonic (normal matter) areas. Dark matter could have played a role in the disk and stars but not its central black hole. There is evidence that the size and brightness of galaxies has not increased in the past several billion years. This contradicts the merger idea that large galaxies grew by mergers.

The Bible has the (black hole) firmament made right after the light and dark matter were created on the second day (time period). It was found that most if not all galaxies have a uniform history of evolution. It begins with a gigantic black hole surrounded by gases collapsing into stars, accretion of matter onto the central black hole, and finally end of accretion and quiescence. The fact is the early Universe was never radiation dominant and very hot. It never reached more than about 10^{13} or 10 trillion degrees due to the prior Universe collapse energy being at about one hundred times too low to make a fireball. This radiation would not disrupt hot gases in orbit around black holes. It would prevent any further growth of the cold dark matter that did not initially coalesce or orbit into black holes. Cold dark shell matter would also be in higher orbits where it coalesced into halos around the black holes (firmament). It came as a great surprise to scientists that all the galactic parameters of circular velocity, light emission, and galactic size were directly related to the size of the central black hole. Photons of light bouncing from masses and being distorted toward the infrared end of spectrum make up the cosmic background radiation (CBR). The gravity of these masses reduces the photon energy as they escape its gravity. The CBR photons only differ by one part in temperature per ten thousand (.0001). from each other. This would limit the size of masses at about 300,000 years after the big bang when the Universe cooled enough to about ten thousand solar masses if they were related. Since scientists use a fireball model, they are not expecting massive black holes millions of times the size of the sun to be in the earliest Universe. Also massive amounts of orbiting hot gases is never in their models. What does happen is these massive orbiting gases obscure evidence of the black holes. The CBR of today was sitting in the core of the big bang matter for eons. This standing caused its Planck spectrum, not galaxies and clusters. Indeed, no one has been able to link the temperature changes in the photons of the CBR to galactic or cluster structures. Starting from a cold big bang shell and hot core gases, galaxy formation is greatly simplified. First the big bang cold dark matter shell laid down a pretty even density of matter for galactic halos. This halo was found at the same density for all galaxies whatever type and luminosity. Black holes formed as the larger dark matter collapsed by gravitation This dark matter pattern was never explained or what follows. After the cold dark matter pattern was set, hot core gases from the core of the big bang followed. The larger black holes captured those gases with the faster velocities. Smaller black holes captured only slower velocity hot gases. Thus protogalaxies were formed with rotating hot gases around black holes and pretty constant density dark matter halos. The larger the galaxies had higher velocity rotation of gases and stars and greater luminosity or light emission. One amazing thing to scientists is galaxies always rotate in the same direction. All galaxies of a given luminosity have the same rotation rates and size. There is no way to explain their rapid rotation rates either which can not be done by collapsing matter alone. They also can't explain the fully formed galaxies that the James Webb telescope has found in the very early Universe. Many of these galaxies have very large black holes. There is no time for them to grow or accrete matter from their fireball and they remain a puzzle.

CHAPTER 13
BLACK HOLES

It seems that black holes or something like them were first imagined by in the late 1700's by the British philosopher John Mitchell and the French mathematician Pierre-Simon LaPlace. LaPlace claimed that they would form according to Newton's Laws of Gravitation if enough matter was present in one place. It would capture light and particle matter and form a 'dark star' with no light able to escape. Mitchell offered similar ideas but without the mathematical foundation of LaPlace. In those days, the particle theory of light prevailed for a number of years. However light was shown to consist of waves shortly afterward by Thomas Young. and this ended talk of dark stars. It wasn't until 1905 did Albert Einstein show light was composed of both particles and waves. No matter what the intensity of light shining on metal atoms, it was only the change in light frequency that induced more electrons to leave a metal plate and flow in a circuit. Ten years later Einstein showed that light must follow the curvatures in spacetime with his formulation of general relativity. Months afterward, Karl Schwarzschild found a different solution to Einstein's equations for a non-rotating 'dark star' . It required only that matter get sufficiently dense to curve spacetime and trap light. Most physicists including Einstein felt the mathematics solution didn't exist although no reason was known to prevent a sufficiently large or dense star from collapsing to infinite density. With the coming of quantum theory in the 1920's, the exploration of subatomic particles began. In 1939 Oppenheimer and Snyder calculated that a star of sufficient size would collapse to a singularity. There was no known way a collapse could avoided one and there certainly is no way to get into a black hole to check for a singularity. The behavior of black holes with the external world can be examined for changes in external gravitation with time. Neutron stars are observable with densities in the nuclear range. The densities here are such that a few cubic centimeters of matter weigh millions of tons. All the space has been squeezed out of atoms leaving the small hard components. Neutron stars are limited to two solar masses and black holes are always over four solar masses. What happens in between is that matter explodes and being unstable makes a strong case for density limitations on matter. Small black holes under four solar masses have never been found. From the 1950's onward, astronomers found areas of space with much high energy radiation. The Hubble space telescope and other larger ground based telescopes identified black holes in the center of most galaxies. It was never clear why these central black holes correlated so well with the size, rotational velocity and light emission of galaxies from a fireball big bang start. Gravitational wave detectors have picked up black hole mergers. Whether black holes have a singularity inside or an area of zero gravity composed on highly squeezed neutrons and protons could not be differentiated by gravitational waves of black hole mergers. The collapse phase of a large mass over four times the size of the sun to singularities was never disputed. Since no one or even light can ever enter the black hole gravitational radius and return to tell us about it, the black hole interior remains a thought process only. Before giving properties of black holes, there is evidence that they are not the permanent structures that the rest of physicists believe. (These are listed more explicitly in my scientific paper at the end of the book). Matter that has been accelerating directly towards out central galactic black hole has suddenly turned away. A nine solar mass black hole had less than one hundredth of the magnetism that a normal black hole should possess According to General Relativity, an electron should be a black hole. Multiple neutron stars with neutron cores just below highly squeezed levels have evidence of ten to 20 percent loss in gravitation. By rights the Big Bang should have been a black hole which we could never escape. The larger the mass collapsing, the less density it has to reach to make a black hole. Sufficient water by itself will form a black hole!

The properties of a stable black hole include the following. There is a gravitational radius also known as Schwarzschild radius is a place where space is bent back on itself so that even light can not escape. Karl Schwarzschild invented the concept of gravity being so strong, a short time after Einstein released General Relativity. They were known as dark stars or frozen stars until Princeton Professor John A. Wheeler invented the name black holes in the 1967-8. The name was banned from use in the widely respected physics magazine Physical Review feeling that the term was obscene. After Wheeler fought a number of battles, the term stuck. However black holes are not black. They release no light, have no reflections, and are invisible as light from behind them is bent around them. They are like the black body which is a physics term for an object that absorbs all light and reflects nothing. They are more like the invisible Biblical firmament (Hebrew: rakea) that Genesis describes. Falling through the horizon of a large black hole would slow time down but you would not be alive long. The difference in gravitation between your head and feet, known as tidal forces would pull you into a long spaghetti- like object. This would be worse near a singularity if it existed. Time dilation does occur. Your heart beat, breathing, metabolism plus any clock with you would run very slow. Light is shifted toward the red end of the spectrum with less energy leaving a gravitational field. The greater the gravitation, the more the wave length is shifted longer and away from the ultraviolet end and towards the red. Visible light runs from shorter wave length violet to longer wave length slower frequency red. The higher the frequency, the more energy light has. Ultraviolet, X-rays and then gamma rays have more energy respectively. Gamma rays are the most light energy known. After infrared, there is microwaves then radio and television have less and less energy with the lowest frequency and longest wave length. All frequency times wave length must total the speed of light which is constant across the entire Universe. There is nothing simultaneous in the Universe since special relativity. One of two 25 year old twins could leave the earth, travel across the Universe at a large fraction of the speed of light and come back to earth. After fifty years of earth time, the twin that remained on earth could be 75 years old. The twin returning from relativistic speeds could be 35 years old. This is known as the twin paradox.

Quantum effects are evident in particles even at the lowest temperatures.

We can start with Helium atoms which are very inert chemically. The outer shell of Helium is filled with electrons. If Helium is at room temperature, the particles are moving around in all directions. If we heat the Helium atoms they will move around even faster and further per second. Now we will cool the particles down past room temperature down toward absolute zero which is about -273 degrees Centigrade or -459 degrees Fahrenheit. Despite the fact that all the heat has been extracted from the Helium, they still have some jittery motion left in them. This much smaller motion is known as zero point motion or simply quantum jitters. Because of Heisenberg's uncertainty principle, the position of a particle times its velocity can not be zero. So even at absolute zero temperature the particles are still moving. What is important here is that quantum jitters do not cause gravitation where as the much larger thermal jitters do. Also highly squeezed particles which have their thermal motion suppressed do not cause gravitation. Although no experiments have been done to support this, there is evidence of this in the Universe. My technical paper at the end has examples of neutron stars and black holes losing gravitational energy.. These remain puzzles to most scientists. To measure gravity in a large mass with only zero point motion would be most difficult with today's technology. Quantum theory maintains that everything has jitters including electromagnetic waves and empty space.

DARK ENERGY

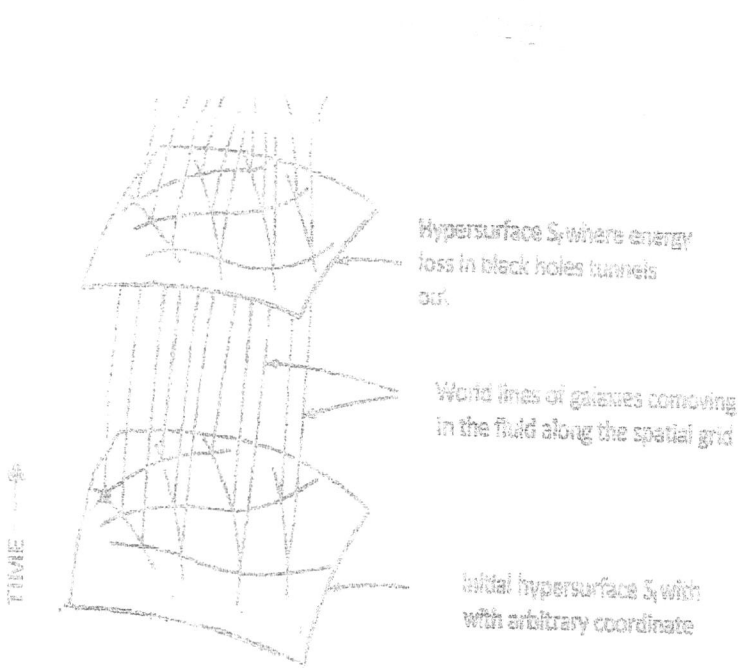

Figure 4: **Galaxy World Lines 3-Geometry.** This a two-space and time dimensional co-moving synchronous coordinate system for galaxies. Shown are two hyper-surfaces with arbitrary imposed grids. If the universe remained homogeneous and isotropic, then proper distances can be calculated. Due to central black hole energy losses, proper galactic distances are increasing.

Professor John A. Wheeler also invented the phrase that 'Black holes have no hair.' By that he meant the black holes have no observable features. In 1972 Jacob Beckenstein was a graduate student of Professor Wheeler. He was interested in how black holes might fit in with Quantum mechanics and thermodynamics. He imagined throwing an amount of hot gas into a black hole. The gas had energy and entropy. When I was in engineering school, we were taught that entropy was a measure of the randomness of object or objects. When you put in heat energy the object(s) become more active with more entropy. Physicists also count entropy as the alternate arrangements of atoms or or information or whatever you are counting. The prior view had been that the gas or whatever you threw into a black hole was gone forever. Beckenstein was able to figure that whatever went into a black hole would add to its mass and energy. In other words, it would increase its size and the size of its gravitational radius. Also the entropy of the gas must add to the black hole. The second law of thermodynamics states that the entropy of any system will increase or possibly stay the same. This law was invented in the days of steam engines and gravitational collapsed objects were never thought of. Beckenstein then made clever use of Heisenberg's uncertainty principle. He thought about adding a single photon of light. Photons of light come in frequencies and wave lengths. When multiplied these two must equal the speed of light. One photon can carry one bit of information. Beckenstein wanted a photon with a wavelength which would be the same size as the gravitational radius. This is also known as the Schwarzschild radius where nothing can escape from including light. So he added the photon to the black hole. Then he used Einstein's famous equation $E = mc^2$ to convert into mass. Then he added it to Schwarzschild's formula to calculate the gravitational radius $Rs = 2MG/c^2$. Here G is Newton's gravitational constant. Finally he used the area of the horizon or the surface of a sphere being $4\pi R^2$. When he finished, Beckenstein found that in one solar mass black hole, the photon increased the mass 10^{-45} kilograms and the radius increased by 10^{-72} meters. The horizon area of a black hole increased by 10^{-70} square meters. These are such an infinitesimal amounts that they are many times greater than that of a proton or neutron. Yet it was real and something. The rule scientists made was adding one bit of information will increase the horizon of any black hole by one Planck unit of area (the increase of the horizon area above). One thing the scientists have left out is that assumes that the highly squeezed matter inside a black hole has the same gravitational energy as normal matter of the same size. In addition, there are no black holes under four solar masses (four times the size of the sun). I will maintain that this is due to a density limitation caused by hard to break down protons and neutrons being highly squeezed, blocking thermal jitters and losing gravitational energy. Going back to standard science, Beckenstein postulated that the entropy of a black hole, measured in bits is proportional to the area of the horizon in (minute) Planck units. In other words information equals area. The problem is there is no easy way to measure the gravitational radius of the surface area it includes. According to General Relativity's equivalence principle of gravity equals acceleration, one wouldn't feel anything falling into the gravitational radius. A person jumping from heights does not feel anything until near the ground or water. It would be like free fall until tidal forces or the difference in gravity between your head and foot begin pulling. So this is a boundary that is impossible to measure exactly from a distance and where no one can return to tell about it from close up. Anyway, the area of the horizon mathematically is proportional to the information and entropy with the two caveats I mentioned above.

In 1974 Stephen Hawking introduced to the world that black holes have a temperature. Hawking had put together quantum mechanics and general relativity. Also he found that small black holes (if any) evaporate. Prior to Hawking it was felt that black holes did not have a temperature. Hawking seemed to get more genius after contracting ALS, a motor neuron disease. Each black hole had a mass and rotation rate calculated by General Relativity. First he started with a physicist's definition of entropy. Entropy is defined in bits of information hidden or not. Then temperature is the increase in energy of a system when you add one bit of entropy. Entropy can be matter or energy. So Hawking saw that from Beckenstein's work one could also calculate the temperature of a black hole. From Quantum Field Theory Hawking knew that electromagnetic fields have quantum jitters. Even in empty space, electromagnetic waves have the jitters called vacuum fluctuations. They are not notable because empty space has no matter or photons to transfer energy. This is much different from thermal jitters. Heating a pot of water, one can see and feel the water swirling around faster and faster due to the heat energy. Quantum Field Theory maintains that even empty space has virtual pairs of photons which get created then quickly reabsorb back into the vacuum. They exist even at absolute zero coming into and out of the vacuum. They are the cause of quantum jitters. Near a black hole horizon, the two type of fluctuations can get mixed up. If object A falls into a large back hole in an area of empty space, there are many virtual photons which are noticeable.

Now consider object B which somehow hovers just above the gravitational radius or horizon despite the tremendous pull of gravitation. There are virtual photon pairs some of which get trapped inside the gravitational radius. Such a virtual photon, with its partner trapped inside the gravitational radius can act like a real thermal photon. So Hawking used Quantum Field Theory to find that these vacuum or quantum fluctuations can lead to a emission of a photon by a black hole. A counter point is that a photon emitted from such an area would lose almost its entire energy climbing out of the hole's gravitational field to a distant observer. This concept was suggested to Hawking in 1972 when he took a trip to Moscow to visit Yakov Zeldovich, the Russian cosmologist. Hawking went on to figure that because of these quantum disturbances, a small black hole would be emitting photons as if it was a hot body. These photons were called Hawking radiation. Hawking further calculated the temperature of the black hole from the entropy and size of the horizon. He found the entropy of the black hole is exactly one quarter of the horizon area in Planck units. The temperature of the black hole was inversely related to the size of the black hole. For a five solar mass black hole this temperature came to 10^{-8} degrees Kelvin or ten billionth of a degree above absolute zero. Small black holes would be white hot with a great number of photons emitted compared to the CBR temperature 2.72 degrees K. However there never have been black holes smaller than four solar masses so there never has been any Hawking radiation found in the Universe. Many physicists have been speculating to what happens to the information in small black holes as they emit radiation and evaporate. I think it is like counting how many unicorns can fit in a phone booth. They don't exist either. Hawking in 1976 postulated that information that fell into a black hole would be destroyed by the singularity. Most physicists felt that information or entropy can never be lost or destroyed, even with the probabilities of quantum physics. In 1981 Hawking argued that information that falls into a black hole are permanently lost to the outside world. Once the information passes the horizon, it would have to exceed the speed of light to escape. He than used Hawking radiation in a small black hole to show the information would be gone. It was like locking something important up in a strong safe then watch the safe with everything in it disappear. In this case the safe contained information or entropy and this caused other physicists look to refute this. This attempt to join General Relativity with Quantum Mechanics led to a conundrum which couldn't easily be resolved.

By using quantum mechanics, Hawking had shown that black holes have a temperature, must emit black body radiation then evaporate. First, there are no small black holes under four solar masses. Number two there are no singularities. All the information in the black hole will be released as the loss of gravity in the highly squeezed core of the black hole spreads slowly outward. This is how the entire information in the black hole will get out as Quantum Theory demands. There will be no gravity one day in the future. The black hole will either explode with much less energy than the big bang. Much energy is lost by matter entering a black hole, about ten percent of rest mass energy. It can simply come apart in pieces of highly squeezed matter in the range of highly squeezed protons and neurons. Small black holes should be white hot as the smallest theoretic black holes would be emitting the most energetic particles. There is a question what these small nonexistent black holes would leave when they evaporate. Would it be the trapped information? As I previously learned in Professor Igor Klebanoff's G.R. class, an electron should be a black hole but isn't. Upon evaporation, any small black hole would leave matter that has no significant gravitational energy such as several tons of normal protons, neutrons and electrons. Baby Universes will not be found in black holes and no one is going one to come out somewhere else in the Universe. All this suffers from pretending matter always has space in it to move and generate gravity ad infinity. Everywhere in the Universe, infinities are followed by zeros which correct it and so it is with singularities. Physicists demonstrate that putting black ink into a large pan hopelessly scatters the ink particles and any information. This doesn't mean that the information in the ink particles is lost. The information is there in the pan but so scattered that no one could put it together. As shown above the information in a black hole will get out and there is not a singularity to destroy it. Although putting energy or mass into a black hole will will increase the horizon, the loss of gravity will decrease the horizon as time marches on. This is how the black hole named the Big Bang happened after all the gravity left during the collapse. Scientists have found no way the Big Bang could have happened as they have no way to cancel the gravity at enormous densities. The information from the Big Bang got out at nuclear densities (squeezed neutrons and protons) just as we are in this world and not at enormous densities.

According to standard science, black holes of a few solar masses would take 10^{68} years to break down into nothing using Hawking Radiation. Since the Universe is only 13.7 billion (13.7×10^9) years, it is obvious that these disappearing black holes never existed then disappeared. According to Hawking, throwing a computer, a car and some food into a black hole would scramble what these represented into a hopeless mess and loss of information. Particle physicists felt that no information was destroyed and everything would be represented on the horizon surface area. It is not clear what the information contained in those objects would be rendered like inside the black hole. Thermal energy always flows from a hotter object to a colder object. The current temperature of the Cosmic Background Radiation (CBR) from the Big Bang is about 2.72 degrees Kelvin about zero. Black hole temperatures from Beckenstein and Hawking are tiny fractions of a degree. So heat flows into the black holes from the surrounding Universe. The radiation temperature drops according to the expansion of the Universe. The theorists extrapolate the supposed dark energy expansion of the Universe sufficiently that all black holes will be hotter than the CBR. Then the black holes will lose energy until they disappear. The major problem with this scenario is that scientists do not know what is causing the supposed acceleration in the size of the Universe. I have posted enough evidence in this paper that there is no acceleration but that galactic black holes are losing energy. The CBR which was finalized at about

375,000 years after the Big Bang is definite that no acceleration of the Universe was occurring. The CBR showed the expansion energy exactly matches the gravitational energy so the Universe was expanding slower and slower. As the central black holes of galaxies lose energy, each galaxy is further and further away than the regular expansion of the Universe would indicate. This dark energy also means that there is some information loss in the black holes. Molecules have been crushed into atoms. Atoms are crushed ito neutrons and protons but not further. When the black holes do finally lose all their gravitation, there will be neutrons breaking down to protons and electrons. Free neutrons will break down under 15 minutes. Their free protons and electrons will combine to produce hydrogen. There will be little to none of the heavier things produced in the Big Bang like deuterium, helium-3, helium-4 and lithium-7. This is due to all the heat and light energy lost in accretion of matter into a black hole. The matter will be there inside the black hole but the information will be scrambled and not recoverable.

There is will little to no scrambling of matter caused by entering the black hole horizon. Certain some scrambling of information is done during accretion of matter into a black hole where matter can lose up to 10% of its rest mass energy. A pair of virtual photons can get one of them across the horizon. The other photon, virtual or real can not hang around the horizon waiting to burn up people or things falling across that barrier. A helicopter hovering in the air has to expend energy against gravity. Photons falling into the ultra strong gravitational field of a black hole would gain tremendous energy, shifting into the blue end of the spectrum then to gamma rays. They would not be able to hover long as the energy losses in an extremely strong gravitational field would lower their energy to harmless radio waves in seconds. A person falling into a black hole would feel nothing except possibly the sting of a few photons that fell in with him. The Equivalence Principle of Einstein where an acceleration is the same as gravity should by and large be more important than quantum effects. So for years Hawking insisted that information would be lost in a black hole singularity despite what the laws of thermodynamics and Quantum theory said. Since virtually all scientists believe that the big bang started at or near a singularity and black holes must have collapsed to such densities of almost infinite density, information must be totally scrambled in such locations. Hawking temperature of large black holes is a very tiny fraction of a degree Kelvin above absolute zero. Photons are small groups of waves which must travel and can not be standing waves without a physical barrier. Any movement in a strong gravitational field would be curved into the horizon. According to Quantum Mechanics, all matter and information is badly scrambled at the Horizon of a black hole. Currently scientists feel there is no solution to conflict between the Equivalence Principle and Quantum Mechanics. However they also feel there is no information loss in black hole evaporation. The information can not get from the black hole interior to the horizon to be released as Hawking Radiation. So they are stuck on the horns of a dilemma. What they don't realize is that black holes are losing gravity. Their core contains particles squeezed together just over the normal density of protons and neutrons. Thermal jitters are suppressed so there is no gravity. Like every other nonexistent infinity in the Universe, singularities are really zero gravity. As this gravity loss propagates out to the surface of the black hole mass, all information in the black hole like the Big Bang will be released to the outside world in a small explosion. The black hole horizon has reduced slowly to zero. Scientists tried to get out of the paradox by imagining small Xerox machines on the black hole horizon copying all the information. One copy falls into the hole and the other remains. The one that falls through the horizon will be destroyed by the supposed singularity. The other copy is thoroughly scrambled by the photons just waiting at the horizon boundary. These photons don't move?

Then it supposedly is radiated back out by Hawking Radiation. However perfect copies of information do not exist. Laws of Uncertainty and Quantum Mechanics are behind this impossibility known as the no cloning principle.

The next thing some physicists cooked up to get out the contradiction between general relativity and quantum mechanics at the black hole horizon was a stretched horizon. It means a thin layer of hot microscopic particles at one Planck distance above the horizon. They imagined this tiny layer in some way similar to the atmosphere around our planet. Every so often, one of the atoms picks up energy and gets ejected into outer space. Losing mass, the black hole must shrink and eventually evaporate. For black holes, there were these contradicting postulates:

1. To an observer outside the black hole, the stretched horizon appears to be a hot layer of horizon atoms that absorb, scramble and later emit as Hawking Radiation every bit of information that falls into the black hole.

2. To a free falling observer, the horizon appears to be empty space. He or she detects nothing special at the horizon although he can not return back past that point without a velocity faster than light. They only encounter destruction later at the singularity.

3. Postulates 1 and 2 are both true but contradictory. The infalling observer was killed at the horizon and survived millions of years inside the black hole.

Before going further into the physicists' black hole story, I will remind you of the solutions. Particles or even photons can not sit at a black hole boundary. It is not a solid boundary on the way in. It is just a boundary in the reverse direction of particles or waves trying escape. Waves move and they and particles could last seconds at a boundary before running out of all possible energy and falling in. The only thing that would bother a free falling observer is the photons falling with him now transformed into gamma rays to sting him. There is no way to follow the dictates of quantum theory that demands the information in a black hole with singularity must get out. There can not be a singularity in a black hole except in mathematical terms. The loss of gravitation in highly squeezed protons and neutrons by losing thermal jitters has been shown in neutron stars and black holes in my attached scientific paper. The information gets out to rhe Universe as large collapsing masses lose gravitation whether they are pre-big bang matter or smaller black holes.

The next thing invented by physicists to get out of the dilemma was called complementarity. Hawking in his lectures maintained that information cannot be radiated out of a black hole. Some felt that there was a fire ring around black hole horizons. The physicists again had three similar possibilities:

1. The information comes out in Hawking radiation.
2. The information is lost
3. The information hides in some tiny black hole remnant. This remnant remains after the black hole evaporated and can be no bigger than Planck mass.

Most physicists felt that the information is lost and a few felt that some tiny remnants are capable of hiding large amounts of information. Finally one said like Sherlock Holmes that after everything is eliminated that is impossible, whatever remains however improbable must be the truth. No one felt like me that all the information will get out as all gravitation will be lost and that black holes are the cause of the unknown dark energy. No physicist questioned the assumptions about singularities. One good engineer was needed. The physicists decided that two contradictory things are both correct about information being lost and surviving into the black hole interior. This is called complementarity. One physicist stated that after dropping one bit of information into a black hole, someone outside could collect the Hawking radiation and recover that bit. He could then jump into the black hole with that bit of information. There would be two copies of the same information in the black hole. It would be a violation of the no quantum Xerox principle. It was a challenge to the complementarity rule. Another physicist suggested they wouldn't meet in the black hole before they crashed into the singularity. Finally they decided it takes an extremely long times for information to be radiated out of a black hole, many zillion times the age of the Universe. By that time the original bit would have hit the singularity and been destroyed. There would never be two copies of the same information there. This is all in the name of getting the information in the black hole out according to quantum theory. According to general relativity, producing enough energy in a small location will produce a small black hole. An electron should be a black hole but is definitely not one. Small black holes if produced by the big bang should not have evaporated by this 13.7 billion year time since the big bang. Working on problems in relativity, like some physicists, I feel that space-time is like a taught rubber sheet. Small masses no matter how energetic can not cause space-time to wrap around itself . It takes more than four solar masses to cause spacetime to wrap itself and trap everything including light. To me this means that matter can not reach greater densities or be packed closer than squeezed neutrons and protons. Obviously massive amounts of water will collapse into a black hole but without any singularity in its core.

The maximum entropy in a region of space-time is one bit per unit area. A hologram is a two dimensional sheet of film of a three dimensional object. They were invented in 1947 by Dennis Gabor, a Hungarian Physicist. Holograms are a type of photograph crossed by strip of interference patterns. Light is scattered off different parts of the object depicted. The photograph is filled with microscopic dark and light patches. Light shined on the scrambling pattern will scatter looking like a floating three dimension object. All the original information is there and can be reconstituted but only to those who know the trick. The image of reality whether galaxies, stars, planets etc can be coded on two dimensions like the black hole insides are represented on the horizon. The required bits of information are at most one bit per Planck area. Where the bit is stored is not critical. Quantum theory states that there is uncertainty in location until the bit is observed. Then its location is determined within a tiny distance. The entropy of a black hole must be greater than the entropy of all the material that went into it. The maximum bits of information that can be packed into a region of space is equal to the Planck pixels or areas that can be packed on the boundary area. This means that there is a representation of everything on the boundary that is inside the black hole. I will add that it presumes that all those highly squeezed particles in the black hole have the same gravity as if they were free to have thermal jiggles. Holograms are quantum images. They can shimmer and flicker like the three dimensional quantum world they represent. Space is a continuum, that is there is no smallest distance. Everything with the smallest distances jitters. Magnetic waves, electric waves even at absolute zero still jitters. At absolute zero they are called zero point fluctuations but still they flutter and jiggle. So physicists have a cut off at Planck length and Planck areas.

In the 1970's the black hole theorists Beckenstein, Hawking and William Unruh showed that near a black hole horizon the thermal and quantum jitters get mixed up. The much weaker quantum jitters get much stronger and the stronger thermal jitters get weaker. The normal quantum jitters are quite dangerous to anything outside the horizon. Space-time is so warped there that the surrounding Universe including stars can appear right above the horizon. As a person and a particle fall through the barrier. they feel nothing except any accelerated photons energized into the gamma ray energies. To an outside observer, things appear more and more slowed at the horizon as the descent of the person and particle seems to stop and never cross the horizon. This is standard science. The complementary (quantum) people believe that the particle is blasted apart by the reversed quantum jitters. Hawking and most black hole theorists in 1993 felt the one could put infinite information in Planck area bits into a black hole. It would be destroyed either by the horizon or later by the singularity. They felt that entropy had nothing to do with counting black hole bits. In order to prove the majority wrong, they postulated that black hole masses can be restricted to certain sizes. This is not the four solar mass lower limit that I've been saying. Rather they said the entropy of black holes has a hidden microscopic substructure. This entropy does not reveal what kind of substructure it represents. They said entropy is proportional to the area of the horizon which is proportional to the square of the Schwarzschild radius which is proportional to mass. Entropy is proportional to mass times mass ($mass^2$). According to this thinking, entropy is not related to mass directly and this bothered the few physicists trying to challenge Hawking. Finally they realized gravity plays a role. As gravity collapses matter, the energy increases in matter but the entropy stays the same. As long as the process is done slowly, the entropy is unchanged. Continue to add gravity as the mass collapses to a black hole. If all the gravity is somehow eliminated, the mass will return to the state it was prior. My cyclical Universe model does this exactly. The gravity is eliminated in highly squeezed neutrons and protons by suppressing their thermal jitters. No one knows if quantum jitters can be eliminated with sufficient squeezing of matter. Examples of black holes and neutron stars losing gravitation are in my paper at the end. Returning to the physics problem and adjusting for gravity, it was found that entropy is proportional to mass times mass.

In 1996 two physicists concocted a black hole with two imaginary things. One was made of strings which are minuscule one dimensional objects smaller than an electron. The second was a D-Brane. This is a two dimensional imaginary object that can stretch and bend. These black holes have zero temperature as they are called extremal where gravitation and electric charge match. What they found as well as other physicists with different types of black holes is that information is never lost. This upholds the second law of thermodynamics that there is no entropy loss in the Universe. The difference I have with these physicists is that the information or entropy gets out like the big bang and not somehow in Hawking radiation. Also black holes lose gravity and violate the second law of thermodynamics as the big bang did.

Figure 4

Figure 5: **Random Motion of Particles**

All matter has random motion. Higher temperature will increase this motion. It is due to quantum effects and causes gravity. When matter is packed very tightly in a black hole this motion is suppressed. This causes black holes to lose gravity slowly causing dark energy.

The scientist that had the most to do with black holes is the founder of black holes Karl Schwarzschild. Schwarzschild was born in 1873 and attended a Jewish school until age 11. He was a child prodigy and had published two papers on orbital mechanics at age 15. He studied for 2 years at University of Strasbourg then later obtained his doctorate at University of Munich in 1896. His doctoral thesis was on rotating bodies and tidal deformations using Poincare's theory. After his dissertation, he worked on methods to determine apparent brightness in stars using photographic plates and how to determine optical density. In 1899 he worked at the University of Munich on photographic magnitudes in different stars. He found that the luminosity was different even at stars the same distance from us. He showed that this was due to surface temperature differences. In 1900 he discussed the possibility that space was non-Euclidean at the German Astronomical society and later that year published a work showing that the radius of curvature of the Universe was greater than 2500 light years. He worked on radiation pressure from the sun and particle size in the tails of comets. He worked at Institute of Gottingen from 1901 to 1909 with David Hilbert and Hermann Minkowski. There he published research on radiation equilibrium in the sun's atmosphere and the transfer of energy through a star. In 1909 he moved to Potsdam as director of the Astrophysical Observatory there. Shortly after Einstein published his work on general relativity, he was surprised to find Karl Schwarzschild had produced an exact solution. At the time Schwarzschild was involved in World War One calculating the trajectories of artillery shells for the German Army. Where Einstein had used rectangular coordinates, Schwarzschild used polar coordinates. Einstein had used approximations of a non rotating spherically symmetrical non charged mass to calculate the precession (advancement) of Mercury as it came closest to the sun. Schwarzschild was able to use exact calculations which he sent to Einstein in a letter December 22, 1915. He noted that even during the war he was able to explore Einstein's ideas.

Schwarzschild coordinates do not contain a point singularity. The Schwarzschild or gravitational radius where no particles or light may escape around a sufficiently large mass is $Rs = 2GM/c^2$. Here G is Newton's gravitational constant, M is the mass of the object and c is the speed of light. Schwarzschild coordinates do break down at $Rs = 2M$ distance from the object and one can not use them closer to a black hole than this distance. Singularities would require particles far smaller than quarks to continue moving to generate gravity. Other coordinates can be followed inside the gravitating mass but there is no singularity as shown in my paper.

CHAPTER 14
STRANGE FIRE FROM ISRAEL

The Bible records that two sons of Aaron the High Priest Nadav and Avihu had offered strange fire that was not commanded by God. They also died in a strange way. The insides of their bodies were burnt while the outside was intact. They had to be removed from the Holy Ark by pulling on their clothes and dragging their bodies outside the camp for burial. I presume this was done so that no one else died. There is also the story of the return of the Ark from the Philistines. It began to fall and a person named Uriah tried to push it back and he died while touching it. If one looks earlier in the book of Exodus, one can find a description of the construction of the Ark. We will use a cubit at 18 inches which is the smaller one used. Acacia wood 2 cubit and a half long by one and half cubit wide and one and 1/2 cubit tall was covered on both sides by gold plates. This is two plates of 6318 square inches. We will assume that the opposite of the Ark of acacia wood was also covered with gold and was 1/2 a inch away. This makes a large plate capacitor of 2.84µ Farads which could easily hold a voltage of 100,000 volts. Now this could supply only direct current. Alternating current could stop the heart at much lower voltages. Just like the Biblical descriptions of Nadav and Abihu, electricity will flow where there is least resistance. This will take place along the blood vessels which have salt water of about 140 milli-equivalents of sodium and about 105 milli-equivalents of chloride per liter. Thus the body could be burnt only on the inside and the two brothers would have to be pulled away from the Ark by their clothes so that no one else was electrocuted. Although this is supposition, it does make physical sense with known properties of high voltage electricity.

Chapter 15
Ageing and The Bible

Ageing as in the Bible may soon be possible so people may live up to 1000 years. People do age due to diabetes, heart disease, hypertension, kidney disease, cancer, immune system decline and cognitive decline. On a cellular level, there is also ageing. As cells replicate, they accumulate errors in the DNA which stores the genes, and RNA which is translated into proteins. As normal cells lose function, stem cells have to step in to replace them. These stem cells also accumulate errors. Even without disease, people still age. In 1962 Leonard Hayflick found that normal human cells would divide about 50 times before they stopped dividing. Various animal cells would divide in some proportion to how long they lived on average. It was later found the ends of the chromosomes called telomeres shorten on every cell division. The reason the cells stopped dividing was traced to two guard proteins called p53 and p16. These proteins stop cell division when the chromosomes get too short or contain too much damage. The stopping of cell division is called senescence. Cell damage that can cause senescence includes radiation such as from neutrinos, oxidation, not repaired DNA damage and so on. Once the guard proteins in the cell are activated, the cell will go into early senescence. This can revert back to normal if the cellular damage is repaired. If the damage is not repaired, the cell will go into full or deep senescence. With full senescence, the cell puts out proteins calling for white blood cells to come and engulf the cell. These proteins also cause decreased wound healing, decreased blood vessel growth and decreased tissue repair in the area where the senescent cell is located. With ageing, the white blood cells don't work well either so more and more senescent cells accumulate in the body. These are the direct cause of ageing. This process was highlighted by an amazing experiment. Researchers operated on old female mice, took out their ovaries and put them into young female mice. They then put the ovaries from the young females back into the old female mice. Subsequently the young females with old ovaries gave birth to mostly normal babies. The old mice with young ovaries gave birth to only few babies with many genetic defects. The senescent cells in aged bodies are the cause of function decline in the entire body. In each organ of the body are stem cells. These cells normally lie dormant in every organ and do not divide. They can be found in the heart muscle, lining the spinal fluid cavities in the brain, in sinusoids of the liver and so on. The more these cells have to exit dormancy and divide to replace normal cells, the more DNA damage they accumulate. Old stem cells as well as senescent cells decrease functioning of every organ they are in. Diseases that cause chronic inflammation such as inflammatory bowel disease, skin inflammation of psoriasis, will cause faster ageing of times. Clearance of senescent cells by white blood cells will reduce inflammation and increase organ function. Mitochondria are the tiny power houses of each cell. Simply put, they process sugars into water, carbon dioxide and energy along with damaging chemicals. In diseases like diabetes, the high sugars cause the mitochondria to work overtime. More dangerous chemicals are produced causing DNA damage and ageing of tissues. High insulin levels, related to the insulin-like growth factor, will cause cells to overgrow in many parts of the body. This leads to blood vessel linings to grow and block up blood vessels and precancerous cells to grow possibly into cancer.

In 1970 a fungus was found that produces the compound rapamycin. This chemical was originally used in immunosuppression for organ transplants. It also kept open stents that bypassed diseased arteries. Subsequently its method of action was found. It shuts down a protein called MTOR, letters meaning 'the mechanistic target of rapamycin.' MTOR senses food requirements and leads to hunger. As ageing was better understood, rapamycin was used to slow metabolism of foods and cause senescent cells to self destruct by a mechanism called apoptosis. This tends to slow or reverse ageing. Another group of scientists developed a drug to specifically eliminate senescent cells. This group found that the guard protein p53 along with the protein FOXO4 maintain senescent cells They developed a protein called FOXO4DR1. This chemical pushed senescent cells into apoptosis and death. Old mice regrew their hair and muscles and ran around like young mice. A number of medicines are being investigated to rid the body of senescent cells.

Reducing food intake will slow ageing as well. The amount of food reduction necessary to slow ageing is 20 percent of normal caloric intake. This type of diet is not feasible for humans. It does markedly reduce insulin and insulin-like grow factors. The diabetic drug metformin will cause a reduction in food intake, metabolism and insulin. It will reduce cancers by several mechanisms and is under study as an anti-ageing compound. Scientists have found another mechanism to slow ageing. MTOR controls food metabolism and protein synthesis. POLIII is an RNA enzyme causing protein synthesis. When this enzyme is blocked and protein synthesis reduced, ageing is slowed. Just eliminating senescent cells will allow the remainder of the cells in the body to multiply up to their Hayflick limits of about fifty times and allow people to live many hundreds of years.

This one key enzyme may have been lost in Biblical days.

CHAPER 16
EINSTEIN'S SPOOKY ACTION AT A DISTANCE

There was a phenomenon in quantum physics that bothered Einstein very much. That is the entanglement of photons or particles. Entangling photons is easy. A laser beam creates photons. They must pass through a wave plate so that the photons are entangled. The beam then strikes a barium borate crystal which splits the beam into two red beams. Each has identical photons being split with identical energies. The split beams go through a polarizer keeping the photons that pass through with the same polarization. The photons are then sent to strong photon detectors which can detect single photons. The detectors are wired to meters so that each photon as it is detected will cause an electric current to flow. A third detector shows when a pair of photons goes through the detectors at the same time. These entangled photons land with the same orientation each time. Photons that are not entangled land with the same orientation only 25 percent of the time. Regardless of the distances between beams, photons will coordinate their results. Researchers have stretched the distance between entangled photons to 100 miles. They tried to measure the speed which one photon knew what was happening to the other. It was at least ten thousand times the speed of light.

Einstein had kept physics local. The infinite speed of Newtonian gravity was overthrown. Special relativity had shown nothing travels faster than the speed of light. General relativity showed that even gravity waves which are distortions of spacetime can only travel at the speed of light. Particles traveling at the speed of light would have infinite mass. Further it was violated cause and effect. It would be like the little saying: Margaret Bright could go faster than the speed of light. She left home one day and came back the previous night. In other words, she came back before she left, which is nonsense. Bohr developed what is called the Copenhagen interpretation from where he lived and started a school. Werner Heisenberg was one of his first pupils. He developed what is called the Heisenberg's Uncertainty Principle. Quantum theory had subatomic particles being only a probability of being in an exact location. There are limits on light wave size and knowledge exactly where in space subatomic particles were, Heisenberg found that it is impossible to know the exact position and velocity of a particle at the same time. This type of indeterminism or probability never bothered Einstein. When quantum theory of Niels Bohr and Werner Heisenberg stated that the wave function of matter could span the entire Universe instantly, Einstein argued strongly against it. Even light which acted both like waves and matter in different places had a finite speed limit. So this wave function, which included one or many particles, could interact anywhere across the Universe instantly. It could bind all the fates of all the particles almost like magic. Because of this non-locality problem, Einstein took the position that it wasn't wrong but incomplete. It violated the relativistic speed limit. He also hated the probabilities. Although Einstein was not religious, he famously said, "God does not play dice." In 1927 Einstein and Bohr met at the first Solvay Conference in Belgium along with twenty seven other famous physicists including Marie Curie of radiation fame. Einstein did not present any papers on relativity but instead argued against the wave function of the Copenhagen Interpretation. During the question and answer period, he wanted to know if a quantum waveform spreads out in space like an expanding soap bubble, how come the particle it represents will appear in space at a specific place. What pops the bubble? What force is operating instantaneously through space? Bohr accepted the non local version as opposed to incomplete which Einstein claimed. Einstein brought up a number of other inconsistencies but Bohr always had an answer about the indeterminate probabilistic nature of space. The next Solvay conference was held 3 years later in 1930. Einstein had another problem for Bohr. A box is filled with photons of light and one escapes. The box recoils in the opposite direction. The box and the particles including the one that escaped are linked by one waveform. You can calculate both the position of the box and that of the photon. The Copenhagen interpretation says both the box and the escaped photon have no specific location until you measure it. After you measure the box, the photon appears. The wave function collapses like magic. Since the photon is going at the speed of light, the wave function must be going even faster than the speed of light to make it appear. Einstein felt that the photon is already where you find it. He stated later that which exists in Part B of space should not depend on what kind of measurement carried out in part A of space. Einstein never had quarrels with the Uncertainty Principle although there were some misinterpretations of what he said. There never was another Solvay Conference as Adolf Hitler came to power in 1933. Solvay and many of the physicists were Jewish, and like Einstein had to flee for their lives. As the Nazis burned his books and claimed Relativity was a hoax, Einstein came to the United States. Einstein moved to the Institute for Advanced Study in Princeton. He got the United States moving in research for the atomic bomb with a letter to Franklin Delano Roosevelt. Two years in the U.S., Einstein along with two junior colleges Boris Podolsky and Nathan Rosen wrote a paper about non locality called the EPR paper. Erwin Schrodinger, who also fled Germany coined the name 'entanglement' for the quantum non locality behavior. He also made the famous scenario of a cat which is neither alive nor dead until someone looks into its cage. There was quite a conflict among physicists after the EPR paper. Bohr was unable to refute most of the EPR paper but fixed his rebuttal on the last part of the paper written by Podolsky. Podolsky tried to stretch Einstein's arguments about entanglement to a refutation of the Uncertainty Principle. Bohr easily disposed that argument but did not refute Einstein's main point about either nonlocality or incompleteness. Most people presumed that Bohr's refutation stood. Further Bohr had a school where he turned out physicists with the Copenhagen Interpretation. Einstein never established a school and his points would have been lost to history if people didn't reexamine his arguments. Even today there is no good rational expression for the phenomena. One answer is that things were rigged from the big bang. Things were stacked in advance. Another answer is that particles are like crystal balls. They knew what was going to happen in advance. The third answer is that there are parallel Universes. This means that they are commanded to match. There are multiple worlds, multiple particles and multiple copies of each observer person. The final non local answer is that the Copenhagen Interpretation is correct and that making a measurement creates reality. Einstein's realistic view that a measurement only records what is there is out. Physicists have conducted experiments to check if Einstein was correct. Simply put, it seems that particles are only real in a fixed place after measurement. Thus there is only spooky action at a distance or entanglement as the most logical of all these illogical arguments. There is no rational alternative. How it works cutting out space and time exactly is unclear. At the first, God created the Universe with containing a highly contracted ball of cold dark matter with a hot core. God said let there be light and God divided the light from the dark. This is all spooky action at a distance and still today is the only physics explanation.

CHAPTER 17
THE UNIVERSE IS CYCLICAL

Our Universe started off about 13.7 billion years ago with a big bang. We know the age of the universe because the Hubble Constant is the rate of expansion of the galaxies. Inverting this number, about 70 kilometers/second/megaparsec gives 13.7 billion years. A megaparsec is 3.2 million light years. We know from the photons in the Cosmic Background Radiation left over from the big bang that the expansion energy exact matched the gravitation energy. This is very strong evidence there was a prior collapse that was stopped before a quark soup or singularity. This collapse energy was imprinted on the matter that stopped the collapse. I maintain that this matter was simply composed of neutrons and protons not broken into quarks. At my neighbor Jefferson Labs, they found that it took 10^{35} pascals or about a trillion-trillion-trillion dynes per square centimeter pressure to break up a proton. As they note, this is more pressure than neutron stars can generate. Unless all the matter is moving at relativistic speed near that of light, a gravitational collapse can not break down neutrons and protons. This can not happen at the beginning of a collapse. It must be even harder to break down these particles when highly squeezed. Also highly squeezed particles lose their thermal jitters and do not generate gravity. This is what stopped the prior Universe collapse and stops black holes from collapsing to singularities. The energy release was quite precise at the big bang. Further the configuration of matter at the start had no gravitation at all. This is known as the flatness problem and the Bible tells us that the big bang had a cold shell and hot core, solving the dark matter problem in the process. There was no fireball which could not match the expansion energy with matter. The universe matter never collapsed to a singularity. The maximum density of matter is about one thousand times that of a neutron star. The Bible explains that the radiation energy was embedded in the core of the Big Bang.

There was no free radiation and just darkness just as the Bible states or gravitation wouldn't balance the expansion. Scientists are insisting that some mysterious force (dark energy) is overcoming gravity and pulling the Universe apart. This is loss of gravitation in super massive gravitational black holes only affecting inter galactic distance inside the Universe and is no way sufficient to obviate the slowing of the Universe expansion. The solution to the flatness problem means that this is enough matter with gravitation so the expansion should be continuing ever more slowly, stop and return. This is much like a ball thrown in the air on earth. The original velocity is transferred to potential energy and then the ball returns. With general relativity, it is the matter and all energy that curves space and time. The more matter, the more curvature. There is no way to know exactly when the expansion stops as the light would be blue shifted much light a train that is coming at us will have a higher pitch. The blue shift on galaxies and clusters coming at us would still take about 15 billion years more to be seen here on earth. As the contraction proceeds a growing super max black hole will engulf matter and eventually all light energy as the Universe's scale factor (you can read radius) drops toward zero. It will all end in a black hole which will slowly lose gravitation as the collapse matter in the core reaches the limiting density, about one hundred times nuclear densities. As the core loses gravitation and the gravitational or Schwarzschild radius shrinks toward zero, matter escapes in a big bang. Scientists say General Relativity breaks down at the big bang. This is not correct. If one assumes a perfect fluid with plenty of room for particles to move, then General Relativity doesn't work above highly squeezed nuclear densities about $10^{17} grams/cm^3$. As most people know, matter takes up space. As density rises and space is squeezed out of the matter there is less and less room for particles to move. As this happens, there is less and less gravity in neutron stars and black holes. Shown here is how dark energy originated. Due to galactic black hole gravity losses, the galaxies are further and further apart. As for the Universe starting and ending in a black hole, this must be true. The way scientists measure entropy is by counting the area of the black hole gravitational radius. The second law of thermodynamics was made in the 1850's when dealing with steam engines. The law says that entropy can increase in processes or remain the same but never decrease. We wouldn't be here if the gravitational radius hadn't shrunk to zero to start the big bang. Even if scientists want to count entropy by using only matter for black holes, the fact is the Universe starting in one and ends in a black hole no matter how they count the entropy. Unless some light energy is lost to outside the Universe at some point, it will be the same black hole time after time. There should be no increase in energy, entropy or anything else.

DARK ENERGY

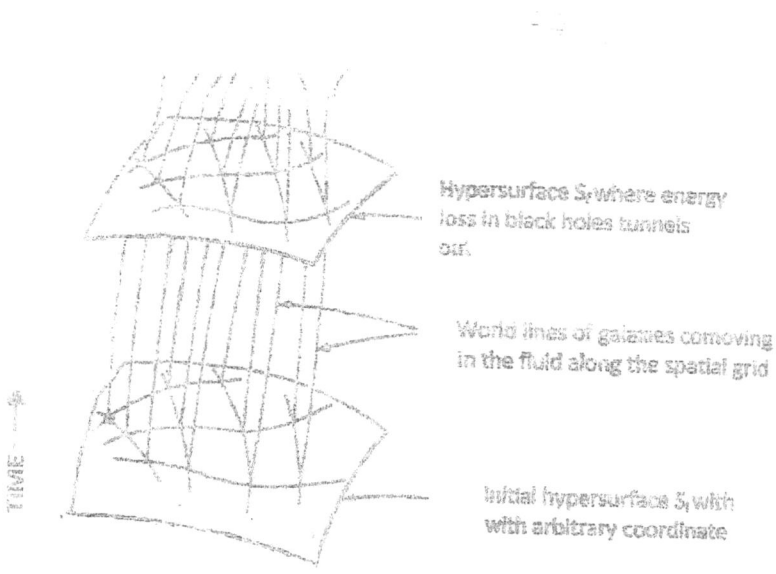

Hypersurface S, where energy loss in black holes tunnels out.

World lines of galaxies comoving in the fluid along the spatial grid

Initial hypersurface S, with with arbitrary coordinate

Figure 6: **Galaxy World Lines 3-Geometry**. This a a two-space and time dimensional co-moving synchronous coordinate system for galaxies. Shown are two hyper-surfaces with arbitrary imposed grids. If the universe remained homogeneous and isotropic, then proper distances can be calculated. Due to central black hole energy losses, proper galactic distances are increased.

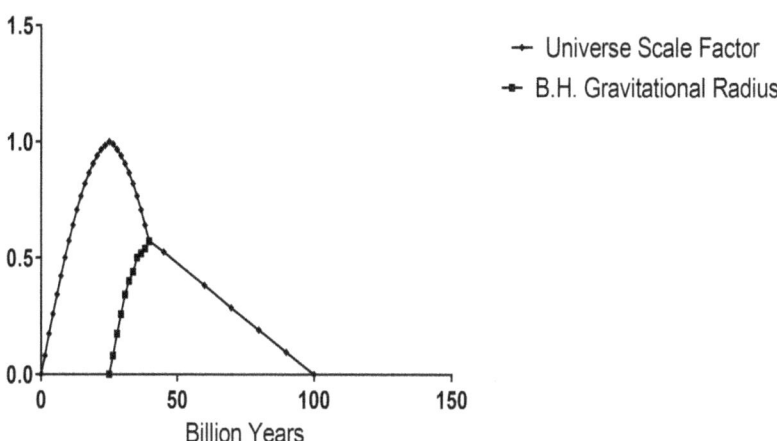

Figure 7: **The Cyclical Universe** started from a spherical shaped mass at or near limiting density. The universe expands to a maximum and then contracts. During contraction, there is a growing super-max black hole. The collapsing universe will eventually force all matter and radiation into its gravitational radius. Kinetic energy of shell particles and radiation will slowly transfer to core. The gravitational energy and gravitational radius will slowly decrease to zero (cycle time estimated). Then core potential energy can start a new cycle of same size. Entropy is not increased due to the crushing of matter back to neutrons by the black hole that preceded the big bang.

CHAPTER 18
SUMMARY BIG BANG SCIENCE VS. BIBLICAL GENESIS

The Biblical Genesis in Hebrew "Maase Bereshit" for the making of the First (Creation in the Universe) solves multiple problems which are included in this work and summarized here.

1. The start of the big bang was not a fireball rather a cold dark matter shell and hot core.
2. This means that with any or all matter including the prior Universe matter or a black hole can not collapse until a singularity. Collapses are stopped by highly squeezed neutrons and protons until they have no gravitation.
3. The Planck Spectrum in Cosmic Background Radiation was caused by standing waves in the hot core prior to the Big Bang.
4. The exact matching of expansion energy and gravitation which produces the flatness in Universe geometry could only come from matter which was able to halt a prior Universe collapse and then release this energy to cause expansion. In physics parlance, this is called a bounce.
5. Galaxies were formed as the cold shell matter collapsed into supermassive black holes. The following hot core gases were captured according to the size of the black holes.
6. The missing dark matter that hundreds of scientists have been unable to find is normal protons, neutrons and electrons from the cold matter big bang shell.
7. Dark energy is caused by large galactic black holes losing gravitational energy as the prior big bang matter did on collapse.
8. There is no way to know how many cycles the Universe went through prior to ours. The black hole formed from the

prior Universe collapse included all light and matter energy. Once the squeezing of matter is released at the big bang, free neutrons break down into protons and electrons which form the primordial hydrogen in less than 15 minutes.

9. Searching for dark matter and writing computer programs to avoid non existent matter singularities, has kept a lot of intelligent men employed. With no evidence that the universe ever was a fireball,. scientists should give up the business of searching for solutions and read the Bible.

10. The Biblical description of the big bang could only have been composed by God. It was then written down by Moses. It must be true not only because God composed it but also the limit on density of matter from the smallest black holes. This means matter can't be packed any tighter. The universe could never have collapsed to a fireball but only to a cold shell of about 10 million kilometers in size and hot core about 10 trillion degrees.. The Bible gives physics lessons too.

The Biblical model actually began in 1981 when I realized that there was an explosion or expansion mechanism in the Bible. The hot core was the propellant of the cold shell. This made more sense to me than the people saying space exploded or not having an expansion mechanism at all or with fireballs so that everything exploded. General Relativity as formulated actually breaks down at the big bang and a fraction of a second after. These other explanations were wild guesses. A cold shell and a hot core allowed an explanation of why gravitation exactly matches the expansion energy. This is called a bounce in astrophysics terminology. I didn't have proof of anything that could stop a gravitational collapse at that point but knew it could solve that flatness problem of why gravitation exactly matched the expansion energy. Once I postulated that neutrons and protons could resist the entire Universe collapse, I realized that this resisting pressure could explain supernovas. Scientists' massive computer programs could not explain why collapsing stars explode. They felt everything would continue to collapse with their soft equations of state. I saw that strong resisting pressure of protons and neutrons could also stop a collapsing star and turn the collapse energy into an explosion.

At the time of the big bang, gravity had to disappear. As an engineer I didn't believe in inflation for many of the reasons I've listed in prior chapters. Once I felt that there was a bounce with a Biblical cold shell and hot core, I began checking whether I could produce galaxies. The Bible had a firmament created on the second day and I began to see if it could solve galaxy formation problems. I found there was a Tully-Fisher relation for spiral galaxies. These scientists had found a good correlation between the luminosity of a galaxy and its rotation velocity. The more light a galaxy produced, the faster it rotated. The light is produced by stars but the rotation rate is related to the total mass including matter not producing light. This relation was found to work even better in the infrared to microwave range compared to visible light. There was an analogous relation for elliptical galaxies, which are those without a central disk, called Faber-Jackson. I realized that the larger the black holes (Hebrew: rakeas) formed, the more gravitation each had to capture the faster hot cores gases later in the big bang. This would explain these correlations and also why the galaxies all rotate in the same direction. With a little help from Prof. John Rollino at Rutgers University Newark, the mathematics fit perfectly and is placed toward the end of my scientific paper at the end of this book. Prof Rollino has unfortunately passed away before this book was written. Scientists using random processes have been unable to explain these two correlations and mostly ignore them. After explaining Tully-Fisher and Faber-Jackson, I knew I was on the right tract. I called Edward (Rocky) Kolb, co-author of the book on the early Universe. He gave me the Fortran language computer program that his student Larry Kawano had written on big bang nucleosynthesis. I talked to him quite a number of times over several years when I started putting astrophysics papers on the internet. The fact the I or rather the Bible had cold matter in the shell to start the big bang was most helpful in to explain the origin of cold dark matter. The cold dark matter formed the halo around galaxies and I later found two articles showing virtually all the galaxies had similar dark matter profiles. The bounce model could explain all the findings including the radiation from the big bang called Cosmic Background Radiation (CBR).

The CBR had a Planck spectrum with variations only about one part in one hundred thousand. Now this could be made by a fireball similar to our sun which has a Planck spectrum. I found that standing waves in a black body cavity could make a similar Planck spectrum. The Bible describes a hot core and a cold shell which would make such a black body cavity where the wall is almost a perfect absorber of radiation. Since no one ever found a direct connection of CBR to galaxies, I added this process to my paper.

It wasn't until 2011 that I found evidence on the internet that astrophysical bodies have lost 10-20 percent of their normal gravity. That year was posted two articles that neutron stars have lost some of their normal gravity. This is the lowest density I could expect if squeezing the particles reduces their motion and gravitation. The core of neutron stars has a density at or just above neutron fluid densities $10^{14-15} gm/cm^3$. Within two years there were articles that objects accelerating toward our Milky Way galactic black hole just suddenly turned away. A nine solar mass black hole had its magnetic field measured. It was about one four-hundreth of the theoretic value. I felt my disbelief of singularities was proven. I added these proofs to my scientific paper. On reading of the history of general relativity (GR), I found the Albert Einstein felt that the stress energy tensor was its weak link. This belief never seemed to stop anyone from extrapolating it from the neutron star densities to 85 orders of magnitude in black holes.

In 2018, I found my neighbors at Jefferson Labs put a most helpful paper in Nature Magazine. It showed that protons had enormous resistive pressures of about 10^{35} Pascals. They also noted that this is more pressure than a neutron star could generate. Although matter collapsing at high relativistic velocities could break down protons, no normal gravitational collapsing matter could reach these speeds. Particles would be unable to move and gravity would be zero in central areas preventing further collapse. Proof of a bounce was found and zero gravity for black hole centers was found.

In 2025 was found further evidence that the Biblical model was correct. The smallest black hole is four times the size of the sun. This mass can not contain any more density or a smaller mass could be a black hole, When this interaction was analyzed, it was found that cold dark matter must consist of highly squeezed neutron/proton sized particles about one thousand times denser than a neutron star. The resultant radiation is about 10 trillion degrees, enough to produce a big bang but about one hundred times too small to produce a fireball. There were no hot quarks for a quark gluon plasma nor antimatter nor any unusual unknown matter.

A short comment about the Exodus from Egypt. In the 1990s, digging at the old city of Ramses in the Nile Delta revealed evidence of the Biblical Joseph and the twelve tribes. The original name of the city was Avaris dating from the Middle Egyptian period. This area was very rainy and the Hebrews called it Goshen from the Hebrew word 'geshem' which means rain.There was a special house built for Joseph and his many colored coat. Later in this time period came the Hyksos, a name invented by the Greeks who couldn't pronounce the Egyptian name for Semitic people from Canaan. They conquered Eqypt and ruled from Northern Egypt from about 1650 to about 1550 BCE. Their capital was in Avaris in the Eastern Nile Delta. They brought chariots, the composite long bow and hardened battle axes when they came. According to their pylons which were later used for New Kingdom building, they were also democratic and there was no god given right to rule in their kingdom. The native Egyptians got word of their planned unity with the Nubians in the South. Kamose based in Thebes began the revolt. After he was killed at Avaris, his younger brother Ahmose cut off Avaris from Canaan and conquered the Hyksos. They left the Nile Delta quite suddenly about 1525 BCE. Ahmose (Egyptian for Child of the moon) was the first Pharaoh of the New Kingdom. He began country wide construction projects and most certainly was a Pharaoh who didn't know Joseph. These projects enslaved the Hebrews. The Bible places the Exodus from Egypt 480 years prior to the first Temple in Jerusalem or about 1440 BCE. This would be during the reign of Hatshepsut, the first female Pharoah. The Ramseside period began about 150 years later when Ramses I, a military man from the Nile Delta took over the throne. His grandson was Ramses II who put his name on many buildings even those of his father Seti I. Avaris was renamed Ramses during his reign and this is where the confusion began over dating of the Exodus. Much of the above history is found in the book "Patterns of Evidence, The Exodus."

CHAPTER 19
THE TEN COMMANDMENTS-AUTHOR'S HEBREW
TRANSLATION

1. I am HASHEM (name of God-not pronounced) your God Who has taken you out of the land of Egypt from the house of slaves. You will not have other gods before Me.
2. You shall not make yourselves a carved idol of any image which is in the heavens above, which is in the land below or the waters below the land. You will not bow down to them and not worship them. For I (name of God) your God am a jealous God who remembers the sins of the fathers on the children to the third and fourth generation of those who hate Me and doing kindness to the thousands that love Me and guard My commandments.
3. Do not take (name of God) your God in vain because (name of God) will not find innocent that who use My name in vain.
4. Guard the Sabbath Day to make it Holy as your were commanded by (name of God) your God. Six days you will work and do all your work. And the seventh day is a rest day to (name of God) your God. You will not do any work: you, your sons, your daughters, your male servants, your female servants, you ox, your donkey, all your animals, your converts which are in your gates in order that your male servant and female servant rest like you. And you will remember that you were a slave in the land of Egypt and your God (name of God) took you out from there with a strong hand, extended arm . On this commanded (name of God) your God about the Sabbath Day.
5. Honor your father and mother as commanded (name of God) your God in order that you will lengthen your days and in order that you will be well on the earth which (name of God) your God gave to you.
6. Do not murder
7. Do not be sexual with a married woman.

8. Do not steal (kidnap)
9. Do not bear false witness against your neighbor.
10. Do not desire your neighbor's wife, your neighbor's house, his field, his male servant, his female servant, his ox, his donkey and all that is your neighbor's.

CHAPTER 20
GLOSSARY

Absolute zero-The lowest temperature that can be reached. Most particles still have a zero point motion of quantum uncertainty effects and thus still moving but much less than with thermal energy.

antigravity-supposed opposite force from gravity attraction, This is a repulsive force.

antimatter-this is a type real matter that annihilates matter when it meets it. A major question is why the big bang didn't produce equal amounts of matter and antimatter in the Universe.

black body radiation-this is radiation emitted from a perfect absorber of radiation. there is no reflection of electromagnetic radiation. It will yield a Planck spectrum.

black hole-a mass in the sky with enough gravitation to bend space back on itself. Nothing can escape in classical general relativity. The big bang as a black hole is unexplained.

black hole complementarity-the use of Bohr's complementarity that light can be particles or waves but not both at the same time. Applied to black holes, it means the matter somehow escapes.

bounce-the stopping of a gravitational collapse by confined particles such that the re-expansion energy matches the gravitational energy.

Brownian motion-The movement of particles floating on water being bombarded by water molecules in motion.

curvature of spacetime. The bending of space onto itself while time is dilated.

dark star-now called a black hole and defined as above.

determinism-the prior to quantum physics idea that everything is determined by matter and forces already existing back to the Universe creation.

electric field-the field of forces surrounding electric charges. electromagnetic waves-traveling disturbances of spacetime by electromagnetic fields like light.

entropy-in Engineering was a measure of disorder of a system. Physicists use it to represent hidden information stored in too many (matter) things to observe.

Equivalence Principle-Einstein used this to show that gravity was just acceleration.

First Law of Thermodynamics-Is the conservation of energy. Energy can be changed but not lost.

Gamma Rays-The highest frequency and shortest wave electromagnetic waves in the Universe.

General Relativity-Einstein's law of gravitation using curved spacetimes. geodesic-the shortest route between two points in curved spacetime. gluonsThe strong force particles that keep quarks together in the nucleus.

ground state-the lowest energy of particles which is absolute zero temperature. There are still quantum jitters for most particles.

hadrons-the basic particles in the nucleus which are gluons and quarks which can be built into larger particles.

Hawking Radiation-Black body radiation emitted by black holes. This was first suggested by Yakov Zeldovitch to Hawking during his trip to Moscow.

Heisenberg Uncertainty Principle-The quantum principle that limits one in determining the exact position and velocity of a particle at the same time. Hertz-The number of electromagnetic waves per second named after Heinrich Hertz.

Horizon-The area from which nothing can escape a black hole. Information-Bits (0 or 1) that determine a state of matter in the Universe.

Infrared radiation-Electromagnetic waves slightly longer than red color which can be used to convey thermal energy.

Interference-The process of adding or subtracting wave energy at different points.

Magnetic field-the force field surrounding magnets and alternating electric currents.

Microwaves-Electromagnetic waves shorter than radio waves which can which can transfer thermal energy and be used for cooking.

Neutron Star-This is is star too far collapsed to form a white dwarf. It is composed of neutrons supported by Fermi Energies. It is not large enough to collapse to a black hole until at least 4 solar masses.

Newton's Constant-The numerical constant in Newtonian gravitation between two masses. It is approximately 6.7×10^{11} in metric units but is difficult to determine precisely.

nucleon-is a proton or neutron

photon-an indivisible quantum of light energy

Planck length-Thought to be the smallest possible length in the Universe $10-33$ centimeters. It is many orders of magnitude smaller than the quarks in particles.

Planck's constant-Relates the energy to the frequency of electromagnetic waves and in other quantum phenomena.

Planck time-The unit of time where General Relativity as currently formulated fails to describe the big bang. In Planck units is 10^{-42} seconds.

Point of no return-The horizon or gravitational radius where particles can no longer escape a black hole.

Proper time-The time that actually elapses on particles moving at relativistic speed. It is solved by the formula $1/1-v^2/c^2$.

Quantum Chromodynamics-Theory describing quarks and gluons and how they make up particles.

Quantum Field Theory-Theory that described particles and wave characteristics of matter.

Quantum Gravity-Theory that unites Quantum Mechanics with General Relativity. Although scientists believe it does not yet exist, this book shows how at just above nuclear densities, gravitation disappears due to squeezing effects dampening particle movement except for quantum jitters.

Radio Waves-Are the longest wave lowest energy waves of electromagnetic energy.

Schwarzschild Radius-The horizon or gravitational radius of a black hole. It was discovered by Karl Schwarzschild two years after General Relativity was formulated.

Second Law of Thermodynamics-Was formulated about entropy increasing or at best staying the same in the age of steam engines.

Simultaneity-The idea that two things could happen at the same time. It was uprooted by special relativity and the speed of light.

Singularity-The fiction that the center of a black hole contains a point of infinite density and gravity. Although infinite pressure can not stop a forming singularity, the loss of gravity due to highly squeezed particles will. An eventual bounce of the big bang and black holes did will occur..

Space-time-The combination of space and time joined by special and general relativity. The speed of light per time is the basis for the space and the time. Things traveling less than the speed of light are called time-like. Things traveling faster than the speed of light are called space-like and do not exist. Telomeres are the ends of the chromosomes which shorten with each cell division. After about 50 cell divisions in humans, guard proteins P53 or P18 stop cell division.

Temperature-measures the increase in energy if entropy is added.

Tidal Forces-The spatial variation of the strength of gravity. Falling into a black hole, the difference of gravity between the head and feet may result in a string-like person.

Tunneling-The quantum phenomena where a particle can pass through a barrier even though it doesn't have the energy to do so classically.

Ultraviolet radiation-electromagnetic radiation with waves just shorter and more energetic than the violet color. It causes tanning, skin damage and will kill most bacteria.

Viscosity-Friction between layers of fluid. Honey is very viscous while neutron superfluids are much less viscous than water.

Wavelength-is the distance between one full wave and another. The shorter the wavelength means the higher the frequency and energy.

World line-is the trajectory of a particle or person in spacetime.

X-rays-Electromagnetic radiation with shorter wavelength and more energy than ultraviolet but longer and less energetic waves than gamma rays.

Zero point motion -is the quantum uncertainty motion at absolute zero after all the thermal energies have been removed.

CHAPTER 21
SPACETIME AND TENSORS

In normal Euclidean geometry space has 3 straight directions in right angles which we can call x,y and z and are known as Cartesian coordinates. To measure a distance traveled in an interval in any direction we will say delta x (Δx or change in x dimension). We there is a change in all 3 directions we say the distance gone is $\Delta x + \Delta y + \Delta z$. However to measure the triangle going from zero to the distance traveled we need to square $\Delta x \times \Delta x + \Delta y \times \Delta y$ and $\Delta z \times \Delta z$. Then we must take the square root. If we use minimally small changes we use dx + dy + dz. To sum these we use ds or a minimally small distance $ds^2 = dx^2 + dy^2 + dz^2$. This is like the Pythagorean theorem for triangles but in 3 dimensions not two. This calculation works well in Euclidean geometry with straight lines. To measure distances we use light speed covering the total distances in the time interval above. To simplify the speed of light will be +1 and the distance of x,y,z be -1 each. In special relativity there is a cube with the diagonal on the left -1. The top row of the diagonal is then -1,0,0,0 where this -1 is for Δx. The zeros mean there is no interaction between dimensions. The second row is 0,-1,0,0. The second term is for Δy with zeros for no interaction with other terms. The third row is 0,0,-1,0. The third term is for Δz with zeros as before. The last row is 0,0,0 +1. The last or bottom row is for the speed of light which we use +1 for it. If we use each of the 4 numbers in the above with sum of the square of the distances $-1\Delta x^2 - 1\Delta y^2 - 1\Delta z^2 + 1c^2$ all in units of light speed, we could see if a certain distance light can reach it. If the distance is exactly what light can reach in the same units, this equation will be zero. If there is more distance than light can reach in the time interval, the equation will be negative due to the sum of negative distances and will be called space-like. If light can reach the sum of the distances, the equation will be positive and realistic. The equation will be called time-like or realistic. Nothing can travel

106

faster than the speed of light. The cube outlined above is actually called the metric tensor. It has diagonal terms of three -1's and +1 on the bottom. Mind the -1 and +1 terms are by convention as some physicists use +1 for space and -1 for light. The most basic way to represent spacetime geometry is the line element. For surfaces the line element dl is the representation of an infinitesimal distance between two neighboring spatial points. This can be represented by an infinitesimal change in local coordinates on the surface. For spacetimes where only time is changing, the spatial distances are zero and the time coordinate $(x^0 = c \times time)$. Whether only time or all spacetime is used, the interval is ds including the time with light speed (=distance) and three dimensional space distance. For two neighboring points in spacetime $ds = g_{\mu v} dx^\mu dx^v$, here the neighboring points μ, $v = 0, 1, 2, 3$. Remember 0 is the time component. $g\mu v$ is called the metric tensor. It used as a tool to measure space and time intervals. It is symmetric by construction so that of the 4 X 4 = 16 components, only 10 are independent.

In 4 dimensional curved spacetime, things are a bit more complicated. We still have the 4 X 4 cube which is called the metric tensor. It has rows of $g11$, $g12$, $g13$, $g14$ for the first row through $g41$, $g42$, $g43$, $g44$ for the fourth or last row. These will obviously reduce to the 1's and 0's above in flat Euclidean spacetime if there is no gravity. Gravity will bend or curve coordinates toward each other and dilate time which has the opposite sign. To keep things simple, we will continue to use Euclidean coordinates here and show how they are curved by gravity. Thus we will avoid polar coordinates, cylindrical coordinates or spherical coordinates. The following things are required to make a metric tensor whether the simple diagonals above of Euclidean space or the more complicated ones involving gravity using the off-diagonal terms. The metric tensor is symmetric. That is $g21 = g12$, $g14 = g41$, $g23 = g32$ and so on. The first bottom number is the row and the second the column. Also taking the determinant of $|g| \neq 0$, that is the determinant of all 16 numbers is never zero. Lastly is that the second partial derivative (see section on differential equations) of the gab are not zero. Thus the elements in the matrix are not just numbers but inter relations of the coordinates. The metric tensor is a fundamental tensor for distance measurement. In curved spacetime, it requires 10 of the $4 \times 4 = 16$ elements in the tensor to measure distances from one point to a neighboring point. Four of the sixteen off diagonal terms are not independent. The great thing about a metric tensor is that it can multiply unlike vectors or tensors and get the same invariant answer regardless of coordinates. The Christoffel symbols Γ are based on changes in three metric tensors. If there are no changes with coordinates (first derivatives) there is no gravity. Symmetry is of great help to find solution to Einstein's equations. There are three rotational coordinates and three more independent coordinates. The expanding/collapsing Universe is much like an expanding balloon. Galaxies are like quarter dollar coins fixed on the surface of the balloon. Thus spacetime in the Universe is homogeneous and isotropic. All the matter in the Universe affects this constant curvature. If there is enough matter and energy, spacetime will curve inward like a sphere. If there is insufficient matter and energy, spacetime will be open like a saddle and the Universe will expand forever. If there is just enough matter to close the universe, it will expand greatly but ever more slowly and then return.

In order to put general relativity on the computer, spacetime must be split up into a three-space metric tensor. Time then is like an arrow piercing curving space sheets. Although not running exactly orthogonal (at right angles) to the curving space sheets, time can be broken very easily into small intervals. A mesh is used to represent the space curvatures. The size of the mesh depends on how many central processing units are in the computer. The multi-processing computer I worked on at LSU had virtually hundreds of separate processing units to represent the space curvatures and so a tiny mesh was used. The program to calculate the space curvatures was called 'Cactus'. What I did not like is the program made great efforts to avoid the non-existent singularities in collapsing black holes. There are dozens of different routines to calculate the space curvature as time progressed. I would have had to rewrite many of them to use zero for the gravitational curvature once nuclear density was reached so that neutron and proton particles could not move. Therefore the singularities avoided on the computer are due to mathematical simulation ones only. In the real world, there is no evidence for singularities. Needless to say, the Relativity group I was in Loni Hyrel 17 did not keep me after 8 months. They appreciated that I had answers for dark matter, dark energy and other big bang problems that no one else had. However my work would overturn everything everybody had done and believed, in computer simulations for the past 40 years.

CHAPTER 22

DIFFERENTIAL EQUATIONS AND METHOD OF LINES FOR SCIENTIST AND ENGINEERS

To discuss differential equations we will start with a person in his automobile. If he traveled at 50 miles per hour for 4 hours he has driven 200 miles. Without realizing it, we have started with a rate of change. 50 miles is a distance and per hour is the time. So velocity is the the rate of change of distance with time $\Delta distance/\Delta time$.. Cars are not always moving. They have to start and stop. This is called acceleration and deceleration. As one starts a car and puts his foot on the gas pedal, the car will accelerate say at 2 miles/hour every second. This will take him 25 seconds to go from zero miles/hour to 50 $\Delta velocity/\Delta time$. Notice when you are accelerating in your car that you are pushed back into the seat. This acceleration is the same as gravity but much weaker. The car acceleration *2miles/hour/sec* $=2.88feet/hour/sec$(2.88/3600) = $.0008feet/sec/sec$. Earth's gravity due its mass distorting spacetime is 32 feet/sec/sec. Gravity is an acceleration where in free fall an object would gain a velocity of 32 feet/sec every second. To convert how far our car went, we would have to convert the acceleration to velocities and sum all the velocities and the time elapsed. This step is like integration which is a sum of all the little velocities to determine distance. We will not pursue integration here but rather see velocity is the rate of change of distance with respected to time $\Delta distance/\Delta time$. Acceleration is the rate of change of velocity with respect to time $\Delta(\Delta distance/\Delta time)/time$. This is why time is written twice when giving car accelerations or gravity. We use d's in place of Δs to express the instantaneous value. Thus the car velocity at any instant would be *d(distance)/dT* and acceleration is $d^2(distance)/dT^2$. To put together the total distance that car went we would have to

110

include the acceleration which is a second derivative distance with time. Also we need the velocity which is a first derivative of distance time. We have a basic differential equation. There are a number ways to solve differential equations which physical science and math people are quite familiar. If we keep the wind friction and road friction of the trip constant, we would have to use partial differential equations. These are what result when keeping some variables like friction constant. They are much more difficult to solve than ordinary differential equations. They result also in gravitation as the 4 X 4 tensors are changed into ten partial differential equations. There are only ten as six are redundant due to symmetry.

Ordinary differential equations have a dx/dy as a first derivative. They may have a d^2x/dy^2 as the second derivative of x. Very often these have analytic solutions like $x = sin(y)$ or $x = e^y$ or some combination of these. In order to solve these on a computer which doesn't understand equations, we have to convert the equation into numbers. We also need initial conditions like when y=0 also x=0. The computer will begin taking a small further interval such that when y=0.1 and will run that through the equation to see what x will be. It will then double y to 0.2 to see if x will be double. If not, the interval is too big and it will half the interval say to .05 and try again. It will finally find the area of equation where it is almost linear such that double the interval will double the result. Then it knows that it is safe to multiply y by the interval and use the 'x' result it gets to go to the next interval. In this way, a curving differential equation can be made into a large number of small straight lines following the curve to within certain limits. There are a number of computer programs that do this, the most common of which is Runge-Kutta.

Partial differential equations are in general more difficult to solve. To handle these, we use the method of lines. A partial differential equation is noted by the term ∂ in front of letters. It can be first or second derivatives and not more in general relativity. This is true because gravity is an acceleration so there are no higher equations in all of general relativity like third derivatives. A simple problem might be the old heat equation $\partial u/\partial t = \partial u^2/\partial x^2$. Here we use u for a dependent variable, despite it looking independent. x and t are independent variables. x is a spatial or boundary variable. Even though our partial differential is one dimensional here, many times this is satisfactory to solve a problem realistically. When general relativity's partial differential equations are placed on the computer, they are divided into 3 spacial equations and time. Time is not quite perpendicular to the space bending and these equations are called 3+1 Relativity. In our equation here, t is an initial value variable. Our equation is considered a parabolic partial differential equation. There are three basic types of partial differential equations. They are elliptical, parabolic and hyperbolic. A simple elliptical equation is $\partial_x^2\phi + \partial_y^2 = \rho$. A hyperbolic equation is of type $\partial_y^2\phi - c^2\partial_x^2\phi = \rho$. To distinguish each of these equations, we will give a combined general equation $A\partial_x^2\phi + 2B\,\partial_x\partial_y\phi + C\,\partial_y^2\phi = \rho$. The coefficients A, B, C are real and differentiable and determine what the equations are. If $AC - B^2 > 0$ then there are coordinate transformations and the equations are elliptical. If $AC - B^2 = 0$ with coordinate transformations, the equations are parabolic. If $AC - B^2 < 0$ then with transformations, the equation is hyperbolic. Only hyperbolic equations have real characteristics. Each equation requires an initial condition. For a system of equations, each equation would require an initial condition at the same value like $t = 0$. A second order equation will require two initial conditions, the second on a boundary. If the partial differential equations can be made into ordinary differential equations, we can do so using Runge-Kutta or similar methods. If we can't solve them with ordinary differential equation solvers, we change the PDEs using the formula $\partial^2 u/\partial x^2 \approx [u_{i+1} - 2u_i + u_{i-1}]/\Delta x^2$ and a remainder term. Here a spatial grid in x is generated over an interval. $\Delta x = x_{i+1} -x_i$ At each point along the grid, the second derivative of u will be defined by an algebraic equation easily handled by a computer. This same algebraic equation can be applied to a first order equation like $\partial u/\partial t$. Thus these grids are known as 'Method of Lines'.

CHAPTER 23
CONSTRUCTING A STRESS ENERGY TENSOR

A t every event in space-time, there exists a stress-energy tensor. Tensors are a collection of related numbers much like a matrix. However tensors transform like tensors which means that changing coordinates will give you the same distances, energies or other result. Stress energy or momentum energy tensors as they are also called, contain information on energy density, momentum density and a stress as measured by all observers at that point. These stresses are due to all forms of matter and all non gravitational fields. This is a symmetric tensor which means it excludes non-symmetric forces such as viscosity, shear stresses and all anisotropic pressures (pressures that don't act in all directions). It does contain a density of mass energy (ρ) and isotropic pressure (P). Its use has been over simplified by the use of an ideal fluid. This is a nonexistent fluid that is incompressible and there is no change in density with pressure. The conservation of energy, momentum and matter are obtained by summing surrounding spacetime. The summing of momentum can be accomplished by using a thin rectangular box shape moving in time. The measured momentum is actually a 4-momentum due the addition of a correction factor for relativistic speeds with the 3 spatial components. The box moves through time and accumulates all the energy including electromagnetic energy and momentum. Like other tensors, the stress-energy tensor contains no coordinates nor reference frames. The 0-0 (row 0 and column 0) or time-time part of the tensor contains the total density of mass energy adjusted by the factor for relativistic velocities *(1/square − root[1 − v^2/c^2])* which is known as the Lorentz factor. The remainder of the 4×4 matrix is a 3×3 matrix for space components of stress and momentum. Electromagnetic energy also adds to the stress-energy tensor in a tensor F. It is composed of functions of the magnetic energy density B and electric energy density E. Because the tensor is symmetrical, there are no preferred directions. The flow of 4-momentum (3 space momentum with relativity time factor) across a three dimensional centimeter with 99.999% space in it and particles are free to move. Thus neutrons and protons will be unable to move at densities around a billion billion grams per centimeter cubed. About one thousand times greater, the quark components will be unable to move. This concept and others of mine has been rejected from many different publishers. Yet there is much evidence that this stress-energy tensor fails. My last scientific paper at the end of the book has multiple examples of unexplained gravitation losses in neutron stars and black holes. Neutron stars lose 10-20 percent gravitational energy. Our galactic black hole had matter going toward it and then moving away. A nine solar mass black hole had almost one thousandth of the magnetism it was theoretically supposed to have. The smallest black hole in the Universe is four solar masses and none has been found smaller. Newtonian gravity uses just density and a gravitational constant G. Scientists have been unable to pinpoint the exact value of this G constant. The stress-energy tensor tells why using mass density and pressure is not sufficient. Temperature and other factors can increase the thermal jiggles, velocities and momentum adding to gravity. The momentum of particles is mass × velocity. Pressure gives some correction for this but is not accurate always as calculated.

With all the matter in the Universe in one place, the mass had to be a super-giant black hole prior to the big bang. The collapse was stopped prior to a singularity as the neutrons and protons could not be packed any tighter than densities mentioned above. They are very hard to break into quarks when squeezed together. Compliments of Jefferson Labs in 2018, it would take far more energy than the 10^{35} pascals or a 1 with 35 zeros or 1.45×10^{34} lbs/square inch to break highly squeezed neutrons and protons into quarks. Thus the center of black holes like the big bang had zero gravitation and the prior collapse was stopped. This imprinted the collapse energy on the matter. As all matter was packed together, the neutrons and protons stopped moving as gravity was reduced to zero. This collapse energy was released at the big bang. After the spirit of God hovered over the matter, the collapse energy was released and rapidly re-expanded our universe. Later the dark energy supposedly expanding the universe is just due to galactic black holes losing gravitational energy. Thus galaxies are further apart then the regular universe expansion would indicate.

CHAPTER 24

INTERNET ASTROPHYSICS PAPERS
BY DAVID E. ROSENBERG, (FORMERLY) LSU RELATIVITY GROUP LONI HYREL 17, BATON ROUGE LA 70808 USA

1. Astro-ph 9810351 A Cyclical Baryonic Big Bang explains the Universe
2. Astro-ph 0012023 Quantum correction for our big bang bouncing universe
3. Astro-ph 0008166 Putting together s cyclic baronic universe
4. Astro-ph 9904320 Galaxy formation with dark matter and dark energy
5. Astro-ph 00003464 Is cold dark matter baryonic? Alternative opinion

From: <prx@aps.org>
Date: Wed, May 7, 2025 at 8:53 AM
Subject: Editorial Acknowledgment XE10909 Rosenberg
To: <drdnrosenberg@gmail.com>

Re: XE10909
    Cyclical baryonic big bang explains the universe

    by David E. Rosenberg

Dear Dr. Rosenberg,

The editors acknowledge receipt of this manuscript on 04 May 2025 and
are considering it as a Research Article in Physical Review X.

It has been indicated at submission that you be designated as the
corresponding author for this manuscript.

If you already have an APS journal account, you will be able to access
this submission by visiting https://authors.aps.org/Submissions and
logging in with your journal account credentials. Alternatively, you
can create one at https://journals.aps.org/signup if you do not have
an account yet.

When sending correspondence regarding this manuscript please refer to
the code number XE10909. This code number may also be used to
obtain current information regarding the status of your manuscript
from our automated Author Status Inquiry System (ASIS) at
https://authors.aps.org/Submissions/status/.

Thank you for previously providing us with your ORCID identifier
(0000-0002-2388-7762).

By providing your identifier, you are permitting us to publish it with
your work and deposit into Crossref, a third-party repository for
publications. We strongly encourage all authors to provide their ORCID
identifier.

Figure 8: **Latest Paper Rejection**

My latest paper explains how the fully formed galaxies found by
the James Webb telescope were formed including the massive black
holes that form the basis for them. It explained also dark matter, dark
energy, why the fireball model doesn't work, what happened to the
missing antimatter. So they thought a second time about publishing
it. They would have to give up a new particle collider, dark matter
searches and all the employed astrophysicists searching for a big
bang solution-see next page.

asas

A Cyclical Baryonic Big Bang Explains the Universe

David E. Rosenberg,
EVMS Norfolk VA. 23507 USA

1 Abstract

Our Universe has multiple examples of unexplained gravitational losses in black holes and neutron stars. The smallest black hole $\approx 4M_\odot$ means the maximum baryonic density is $\rho \approx 10^{17} grams/cm^3$. Any collapse of the universe will stop with a scale factor $\approx 10^{13}$ cm. and radiation energy ≈ 10 GeV. Due to higher squeezed core baryons, the outer part of the mass transferred heat energy to the core and became cold dark matter. After contraction reduced particle motion and gravitation, the core radiation energy propelled pieces of the shell into the universe. Each of these masses captured hot core gases according to its gravitational size, forming proto-galaxies. A cold shell and a hot core will explain the Planck spectrum and large galaxy formation in the early universe. Thus the universe was never radiation dominant. The universe will remain cyclical as any increase in the entropy of matter will be crushed back to neutrons during the contraction phase.

Key words: Cosmology theory, dark matter, dark energy, galaxy formation

2 Introduction

There are a number of unsolved problems with the current hot big bang model of the Universe. General relativity has been verified in the Universe except in extreme density conditions: the big bang and black holes. Galaxies are constructed similarly despite their origins being physically too fair apart to be in casual contact (the horizon problem). Initial spacetime was Minkowskian-the big bang expansion energy exactly matched the gravitational energy (the flatness problem). The massive pre-big bang matter must have been a black hole. There are large masses 9 and $10 \times 10^{10} M_\odot$, unexplained in the early Universe. The hot synthesis of light elements occurred in only $4-5\%$ of the matter present. Extrapolation of general relativity many orders of magnitude in the big bang and black holes to or near singularities has not been successful in solving the Universe problems. There is no evidence that the Universe ever reached Planck energies. In the 'Road to Relativity' Einstein felt the gravitational tensors were 'rock solid' but stress-energy tensors were made of 'wood'. The strange physics of highly squeezed matter will be explored here.

Friedman-Robertson Walker geometry generates an isotropic homogeneous spacetime which prevents an initial big bang picture..

$$c^2 d\tau = c^2 dT - a(t)^2 d\Sigma^2 \tag{1}$$

where

$$d\Sigma^2 = \frac{dr^2}{1 - kr^2} \tag{2}$$

where $k = -1, 1, +1$ depending on whether the universe is open geometry, flat or closed.

Neutron stars are presumed to have crusts with a surface region densities $< 10^{14} g/cm^3$. Here in beta equilibrium, there are electrons, neutrons, and nuclei. Relativistic degenerate electrons comprise most of the pressure. The baryon density is near nuclear saturation density $n_0 \approx 0.16\, fm^3$. The general presumption has been that the core should be contain even more relativistic gravitational matter. Applying general relativity to a static spherical symmetric metric gives

$$ds^2 = -e^{2\Phi(r)} c^2 dt^2 + e^{2\Lambda(r)} dr^2 + r^2 d\theta^2 + r^2 sin^{(2)}\theta\, d\phi^2 \tag{3}$$

where ϕ is the azimuthal angle, θ is the polar angle and the radial coordinate r is defined such that at the origin the circumference of a circle is $2\pi r$. Using a Schwarzschild geometry at the star surface, the boundary condition mandates

$$\Phi(r = R) = \frac{1}{2} ln\left(1 - \frac{2GM}{Rc^2}\right) \tag{4}$$

$$\Lambda(r = R) = -\frac{1}{2} ln\left(1 - \frac{2GM}{Rc^2}\right) \tag{5}$$

If one assumes a perfect fluid which oversimplifies calculations, this leads to the Tolman-Oppenheimer-Volkov (TOV) equations:

$$\frac{dP}{dr} = -\frac{Gm\rho}{r^2}\left(1 + \frac{P}{\rho c^2}\right)\left(1 + \frac{4\pi r^3 P}{mc^2}\right)\left(1 - \frac{2Gm}{rc^2}\right)^{-1} \tag{6}$$

$$\frac{dm}{dr} = 4\pi r^2 \rho \tag{7}$$

Here $P = P(r)$, $\rho = \rho(r)$ and the mass within the radius r is $m(r)$. Since the TOV equations are relativistic, the pressure P will add to gravitation especially at the core. At this boundary condition, $m(0) = 0$, $\rho(0) = \rho_c$. With general relativity the pressure terms in the TOV equations will add much gravitation. However matter is not infinitely compressible. It has been found that squeezing protons to 0.3 fm. (femtometer) yields an enormous resistive pressure of 10^{35} pascals[4]. Squeezing protons to 0.1 fm will cause them to become highly repulsive. As they note, this requires more pressure than a neutron star core can generate. The following examples are going to show that gravitation is reduced both in neutron star cores and black holes.

An unreasonable effectiveness of the post-Newtonian approximation has been found in strong gravitational fields of neutron stars[28]. The post Newtonian approximation assumes that gravitational fields in and around bodies are weak and the motions of matter are slow compared to the speed of light. Thus

$$(v/c)^2 \sim GM/\tau c^2 \sim p/\rho c^2 << 0.1 \tag{8}$$

where v, M and τ are velocity, mass and separation of the system. Within masses, p and ρ are the pressure and density and G and c are Newton's gravitational constant and the speed of light. The approximation in post-Newtonian calculation is the 'reduced' Einstein equation

$$(-\partial^2/\partial(ct)^2 + \nabla^2)h^{\alpha\beta} = -16\pi(G/c^4)\tau^{\alpha\beta} \tag{9}$$

where $h^{\alpha\beta}$ is the deviation of the space-time metric $g_{\alpha\beta}$ from the flat Minkowski space-time metric $\eta_{\alpha\beta}$. They are related by the equation

$$h^{\alpha\beta} \equiv \eta^{\alpha\beta} - (-g)^{1/2}g^{\alpha\beta}. \tag{10}$$

If gravity is weak here, then the gravitational tensor potential $h^{\alpha\beta}$ must be small and can approximate the highly nonlinear equations of general relativity.

In 1974 a binary pulsar PSR 1913 + 16 was observed. Both neutron stars had masses $\approx 1.4 M_\odot$ in a quite relativistic orbital system with a mean speed $\approx 200\, km/sec$. Unexpectedly, it was found that the rate of decay of the orbit was in agreement with the post-Newtonian quadrupole formula. More recently, the relativistic double pulsar $J\,0737 - 3039$ performed according to the post-Newtonian calculations despite being in very strong gravitational fields. An effacement process had been postulated that the gravitational binding energy reduces the gravitational mass of each pulsar by $10 - 20\%$ compared to its rest mass.

Another binary of low mass, AXJ1745.62901 has continuously been losing extra orbital momentum over a 20 year period[18]. The high rate of orbital period decrease is $\dot{P}_{orb} = -4.03 \pm 0.32 \times 10^{-11} s/s$. This is over an order of magnitude greater than expected loses due to gravitational waves, magnetic breaking, or conservative mass transfer. The path also included unexplained 'jitter' of $10 - 20$ seconds advancing or retarding the orbital period. Orbital loss explanations from accretion, a third body mass or an unrealistic $.001 M_\odot$ in the outer disc were rejected.

Black holes will form at mass densities

$$\rho = (c^6)/(G^3 M^2) \tag{11}$$

where the constants $c^6/G^3 = 6.272 \times 10^{84} grams^3/cm^3$. The smallest black holes found in today's Universe are $4 M_\odot$ and will form at the highest squeezed baryonic matter $\approx 10^{17} grams/cm^3$. Smaller black holes requiring higher densities to form are nonexistent. Best estimates of the total universe mass are $\approx 1.5 \times 10^{56}$ grams. After this collapsed, a mass with scale factor $\approx 10^{13}$ cm. will result with a radiation energy ≈ 10 GeV. While this is enough energy to propel galaxies a good fraction of light speed, it is over a hundred tines too little to change baryons to a quark-gluon plasma. Highly squeezed baryons should require even more energy.

A unexplained survivor of an encounter with our Milky Way Super-massive Black Hole has been found[2]. From 2006 a G2 'cloud' was found accelerating toward our galactic center mass $\approx 3 \times 10^6\, M_\odot$. The closest approach took place about February 2014. By Sept. 2014 G2 cloud was mysteriously moving away from the black hole. It even failed to trigger a flare up of accretion activity.

A strangely weak magnetic field of a $9 M_\odot$ black hole in the binary V404 Cygni system[6]. A burst of radiation was studied from a flare. It contained charged particles with electrons and protons in the black hole magnetic field. By calculating how quickly the burst dimmed, a team of astronomers found the magnetic field to be 461 ± 12 gauss, the strength of several bar magnets and almost 3 orders of magnitude less than theory predicted.

A galaxy has been found which should not exist according to classical general relativity[3]. Galaxy NGC 3147 has a mildly relativistic Broad Line Region circling and very close $< 100\, r_*$ to the central black hole in an $L/L_{Edel} \approx 10^{-4}$ system. This contradicts the current understanding of accretion flow configuration at extremely low accretion rates. The only reason matter could be closely circling and not accreted in a massive galactic black hole is that the hole must have lost much of its gravitation.

3 The Case Against Singularities

Using general relativity, density infinities have been extrapolated in black holes and where the big bang started. If the universe started at or near a singularity and expanded, there should still be evidence of many of the following missing high energy phenomena. Monopoles are formed $> 10^{14}$ GeV. Antimatter is formed > 38 MeV. Domain walls would be the size of $10^{28}h^{-1}$ cm. and have a mass $\sim 4 \times 10^{65}\Lambda^{1/2}(\sigma/100 GeV)^3$ grams or many orders of magnitude over the present Hubble volume. Its presence would cause a large defect in the CBR. Assuming the universe started as a fireball, the production of geometric flatness $\left(a_{,t}/a\right)^2 = -k/a^2 + \Lambda/3 + 8\pi\rho_m a_o^3/3a^3 + 8\pi\rho_r a_o^4/3a^4$ with $k = 0$ has required an unproven inflaton scalar ϕ, leaving most other problems unsolved. Despite multiple attempts to find evidence of inflation, no evidence has been found in the CBR polarization or anywhere else. Fireballs can't make high correlations in galaxies with the velocity-brightness relation, corresponding rotations and all galactic parameters related to the central black hole size(see below). Considering the small size of the Higgs Boson (125 GeV), the big bang could not produce any cold dark nonbaryonic matter. Nucleosynthesis (the highest big bang energy confirmed) occurred after the end of quark-hadron boundary, requiring $\approx 95\%$ cold dark matter.

4 Explaining Gravitation Loses

Gravitational masses have been treated as pinpoint sources. However, it is not logical that the gravitational properties of an infinitely collapsing highly squeezed mass would be the same as normal matter. For gravitational losses, matter must be highly squeezed and not a gas of noninteracting particles (perfect fluid) nor a Quark-Gluon Plasma. Nucleons highly squeezed to 0.3 fm. must be packed so tightly by $10^{17} gm/cm^3$ that there is no more remaining space nor motion except quantum jitters at all. In order to construct a stress-energy tensor [14], the matter 4-velocity $\mathbf{u} = (dt, 0, 0, 0)$, possibly adding only the observer 4-velocity. The 4-momentum is

$$\mathbf{p} = m\mathbf{u} = (m\gamma, m\Delta x\gamma, m\Delta y\gamma, m\Delta z\gamma) \qquad (12)$$

where $E = m\gamma$, $\gamma = 1$ nonrelativistic and $\Delta x, \Delta y, \Delta z = 0$. A volume element Σ composed of basis vectors e_x, e_y, e_z have zero magnitudes as there is no flow of 4-momentum. An observer in his Lorentz frame will measure mass-energy density in gm/cm^3 $T_{00} = T(e_0, e_0)$, with the observers 4-velocity $\mathbf{u}$ replaced by e_0. If the box is at rest in the observers frame, all matter kinetic and quantum energy will be sufficiently damped except possibly quantum spin. Space-time interaction will cease, $T_{00} = 0$. To construct a volume in spacetime with a parallelopipid, use four different vectors for edges $\mathbf{A}, \mathbf{B}, \mathbf{C}, \mathbf{D}$. The vectors in standard Lorentz frame are $\mathbf{A} = (\Delta t, 0, 0, 0)$, $\mathbf{B} = (0, \Delta x, 0, 0)$, $\mathbf{C} = (0, 0, \Delta y, 0)$ and $\mathbf{D} = (0, 0, 0, \Delta z)$. A 4-volume is

$$\Omega = \epsilon_{\alpha\beta\gamma\delta} A^\alpha B^\beta C^\gamma D^\delta = \mathbf{A} \wedge \mathbf{B} \wedge \mathbf{C} \wedge \mathbf{D} \qquad (13)$$

A volume integral of a tensor $\mathbf{S}$ defined over a four dimensional region $\mathcal{V}$ of spacetime, calculated in a Lorentz frame

$$M^\alpha_{\beta\alpha} = \int S^\alpha_{\beta\gamma} dt\, dx\, dy\, dz \qquad (14)$$

The energy density measured in such a volume $E = m\gamma/V = 0$ as is the density of the 4-momentum $d\mathbf{p}/dV = 0$ per 3 dimensional volume in an observers Lorentz frame. In the following, $j, k = (1, 2, or\, 3)$ in what really is a symmetric tensor. $T^{j,0} = 0$ is the momentum density, j component. $T^{0,k} = 0$ is the energy flux, k component. $T^{j,k} = 0$ is the j component of force from matter and fields acting around x^k. This keeps the tensor divergence $\nabla \cdot \mathbf{T} = 0$ as there is no particle movement.

With rotating immobile squeezed nucleons, let S be a spacelike hypersurface with arbitrary event $\mathcal{A}$ and coordinates $x^\alpha(\mathcal{A}) \equiv a^\alpha$ using globally inertial coordinates. Total angular momentum on S about $\mathcal{A}$ can be defined as

$$J^{\mu\nu} \equiv \int_S \mathcal{J}^{\mu\nu\alpha} d^3 \sum_\alpha \tag{15}$$

and will add to total momentum only if present. Here

$$\mathcal{J}^{\mu\nu\alpha} \equiv (x^\mu - a^\mu)T^{\nu\alpha} - (x^\nu - \mathcal{A}^\nu)T^{\mu\alpha} \tag{16}$$

If S is a hypersurface of constant time t then

$$J^{\mu\nu} = \int \mathcal{J}^{\mu\nu 0} dx\, dy\, dz \tag{17}$$

In the systems rest frame, let $P^0 = M$, $P^j = 0$ and at the center of mass

$$x_{cm}^j = \frac{1}{M} \int x^j T^{00} d^3 x. \tag{18}$$

For a large mass, intrinsic angular momentum is defined angular momentum about any event (a^0, x_{cm}^j) on the world line of the center of mass. Here components $S^{0j} = 0$ and

$$S^{jk} = \epsilon^{jkl} S^l. \text{ and } S \equiv \int (x - x_{cm}) \times d\mathbf{p}/dV \, S^{\mu\nu} d^3 x \tag{19}$$

The intrinsic angular momentum 4-vector S^μ has components in the rest frame (0,S)

$$S^{\mu\nu} = U_\alpha S_\beta \epsilon^{\alpha\beta\mu\nu} \tag{20}$$

The 4-velocity center of a large highly squeezed mass $\mathbf{U}_\beta \equiv \mathbf{P}_\beta/M = 0$. Angular momentum is composed of intrinsic and orbital parts. An arbitrary event a whose perpendicular distance from the center of mass world line is $-Y^\alpha$ making $\mathbf{U}_\beta Y^\beta = 0$. The total angular momentum $J^{\mu\nu}$ about $\mathcal{A}$ is both the intrinsic part

$$(S^{\mu\nu} = \mathbf{U}_\alpha \mathbf{S}_\beta \epsilon^{\alpha\beta\mu\nu}) \tag{21}$$

and the orbital part

$$(L^{\mu\nu} = Y^\mu P^\nu - Y^\nu P^\mu) \tag{22}$$

With the angular momentum about $\mathcal{A}$ and the 4-momentum known (zero is this case) one can calculate the vector from $\mathcal{A}$ to the center of mass world line.

$$Y^\mu = -J^{\mu\nu} P_\nu / M^2 \qquad (23)$$

Using a swarm of identical particles with event $\mathcal{P}$ inside the swarm, m_A is the rest mass. $\mathbf{u}_A$ is the 4-velocity, and $\mathbf{p}_A$ is the 4-momentum. N_A is the number of particles per unit volume, as measured in the particles own rest frame. The number flux vector

$$\mathbf{S}_A \equiv N_A \mathbf{u}_A \qquad (24)$$

The particles have ordinary velocity v_A, zero in packed supranuclear densities. $\mathbf{u}_A^o$ is the the Lorentz correction for volume and velocity $1/(1 - v_A)^{1/2}$. The 4-momentum density is

$$\mathbf{p}_A S_A^o = m_A u^u N_A u_A^o \qquad (25)$$

Consequently the 4-momentum density has components

$$T_A^{uo} = p_A^u S_A^o = m_A N_A u_A^u u_A^o \qquad (26)$$

The flux of the μ component of momentum with perpendicular projection e_j is

$$T_A^{\mu j} = p_A^\mu S_A^j = m_A N_A u_A^\mu u_A^j \qquad (27)$$

Here superscripts (μ, o) and (μ, j) of the frame independent equation

$$\mathbf{T}_A = m_A N_A \mathbf{u}_A \otimes \mathbf{u}_A = \mathbf{p}_A \otimes \mathbf{S}_A \qquad (28)$$

By summing over all categories, the total number flux vector and stress energy tensor are obtained for all particles in the swarm. If $\mathbf{u}_A = 0$, these will be zero.

$$\mathbf{S} = \sum_A N_A \mathbf{u}_A \text{ and } \mathbf{T} = \sum_A m_A N_A \mathbf{u}_A \otimes \mathbf{u}_A = \sum_A \mathbf{p}_A \otimes \mathbf{S}_A \qquad (29)$$

The total momentum flux accross a closed 3-dimensional surface must vanish $\oint T^{\mu 0} d^3 \sigma_a = 0$. There is no flux and no sinks and there is no momentum at these supranuclear densities.

Thorne has a conjecture that a black hole may form only when a given amount of mass-energy collapses through its own Schwarzschild radius $R_S =$

$2M$, thus achieving compactness $M/R > 0.5$[25]. A sufficient large collapsing shell is necessary. If one starts with collapsing black hole nucleons, each weighing $\approx 1.672 \times 10^{-24}$ gm., maximally squeezed to 0.3 fm radius and each with internal repulsive pressures $\sim 10^{35}$ pascals, they can not be packed to more than $\approx 10^{16} gm/cm^3$ density. Due to complete core gravitational loss with thermal jitters and possibly zero point motion suppressed (use stress-energy tensor formation above), collapsing matter can not overcome these internal pressures. This is the core of black holes. Not near Planck energies. Like neutron star cores, black holes will begin to lose gravitation by $\approx 5 \times 10^{14} gm/cm^3$. Once the core begins to form, its little to no gravitational energy and virtual non-compressibility blocks further gravitational collapse. It takes a total of $4M_\odot$, including the non-gravitational core to produce a black hole. The presence of a nongravitational core will reduce $M_{gravitation}$ but otherwise not affect gravitational waves in black hole-black hole coalescence.

Gravitational losses in neutron stars must begin in their cores with density ranges $10^{14-15} gm/cm^3$. With a normal stressed medium, thermal jitters suppressed, having velocities $|\mathbf{v}| << 1$ with respect to a specific Lorentz frame, the spacial components of the momentum are $T^{0j} = \sum m^{jk} v^k$. Here

$$m^{jk} = T^{\bar{0}\bar{0}} \delta^{jk} + T^{\bar{j}\bar{k}} \tag{30}$$

Here $T^{\bar{\mu}\bar{\nu}}$ are components of the stress energy tensor in the rest frame of the medium. Inside a neutron star core, where velocities are known low, and probably zero $T^{\bar{0}\bar{0}} = 0 \sim T^{\bar{j}\bar{k}}$.

How to produce dark matter. After a gravitational collapse is halted by high squeezing of neutrons, the most highly squeezed particles in the core will act as an energy sink. Geometrized units will be employed such that $c = 1 = G$. From the second law of thermodynamics

$$T\,ds = d(\rho V) + p dV = d[V(\rho + p)] - V dp \tag{31}$$

here s is the entropy of the matter and $V \propto a^3$ is the co-moving volume. Integrating gives $dp = (\rho + p)/T\,dT$. Substituting this into the above gives two equations,

$$d/dt\Big[(V(\rho + p))/(T)\Big] = (VE_s)/(T) \tag{32}$$

and

$$\int ds = \int d\Big[(V(\rho + p))/(T)\Big] \tag{33}$$

126

The shell entropy change is $\dot{s}_2 = E_s V/T_2$ and the core is $\dot{s}_1 = -E_s V/T_1$. Normally the energy and entropy increment would follow the temperature differential to a lower temperature as with radiation and dust or with a scalar field and radiation.

$$\dot{s}_{total} = \dot{s}_2 + \dot{s}_1 = E_s V \left(1/T_2 - 1/T_1\right) \tag{34}$$

Quantum effects cause some strange properties. Temperature is a function of the velocity or kinetic energies of the particles, $T^\circ K = m\bar{v}^2/3$.

$$\int d\rho = \int ((\rho + p)/(n)\, dn - nT\, ds) \tag{35}$$

Due to more restricted motion of core baryons, shell energy will be transferred to the core. Core particles will have increased potential energy of deformation but further restricted movement reducing entropy and kinetic energy. Thus cold dark matter originated in shell baryons. The big bang hot core was about $4 - 5\%$ of the total mass. There is still baryon conservation $dn/d\tau = -n\nabla \cdot u$. Energy conservation will still occur. Only $d(nsV)/d\tau \leq 0$. Quantum gravity effects will put matter in better order, that is with less entropy and kinetic energy.

Two teams of astronomers published work in $1997 - 1998$ trying to determine the geometry of the Universe by using Supernova Type 1A as standard candles[17],[19]. They found that the supernova were $10 - 15\%$ further away than even a low density Universe $\Omega_M = 0.2$. Some negative gravitation or dark energy was canceling all the universe gravitational mass-energy. A black hole energy loss will affect the measured supernova distances of co-moving galaxies, as shown in Fig. 1. Spacetime geometry $d\sigma^2 = g_{ij}(t, x^t)dx^i dx^j$ of each hyperspace is assumed to be the same due to homogeneity of the universe. The initial hyper-surface S_I 3-geometry is $\gamma_{ij}x^K \equiv g_{ij}(t_I, x^K)$. At time t_I on surface S_I, they are separated by the proper distance

$$\Delta\sigma(t_I) = (\gamma_{i}, \Delta x^i \Delta x^j)^{1/2}. \tag{36}$$

At some later time t_f, they will be separated by some other proper distance $\Delta\sigma(t_f)$. When spacetime is isotropic, then the ratio $\Delta\sigma(t_f)/\Delta\sigma(t_I)$ will be related to the Universe scale factor $a(t_f)$ at time t_f. The Wilkinson Microwave Anisotropy probe combined with the Hubble Space Telescope resulted in a very small value for the cosmological constant $\Lambda = 3.73 \times 10^{-56} cm^{-2}$

DARK ENERGY

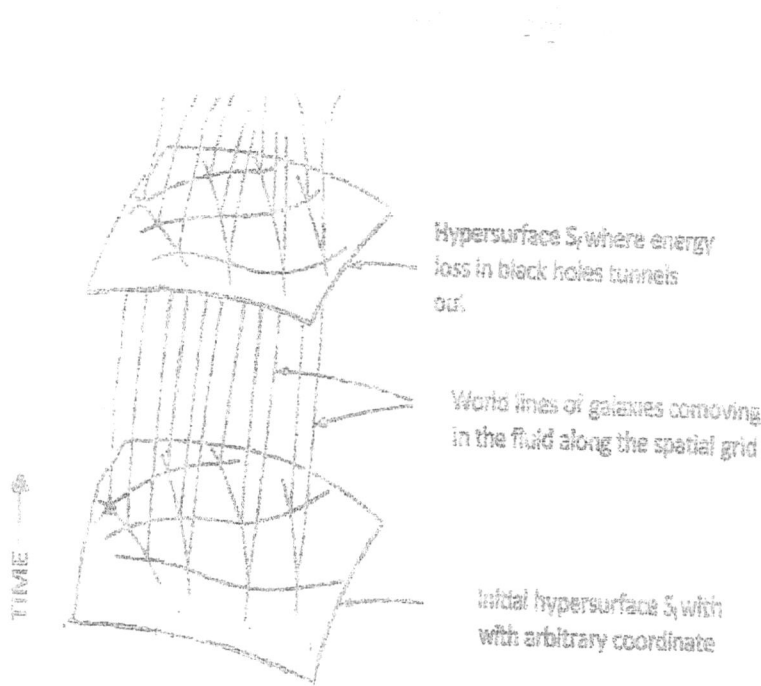

Hypersurface S, where energy loss in black holes tunnels out.

World lines of galaxies comoving in the fluid along the spatial grid

Initial hypersurface S, with with arbitrary coordinate

TIME →

Figure 1: **Galaxy World Lines 3-Geometry.** This a two-space and time dimensional co-moving synchronous coordinate system for galaxies. Shown are two hyper-surfaces with arbitrary imposed grids. If the universe remained homogeneous and isotropic, then proper distances can be calculated. Due to central black hole energy losses, proper galactic distances are increasing.

which corresponds to $\Omega_\Lambda = (\Lambda)/(3H_0^2) = 0.721 \pm .015[11]$. Here $H_0 = 70.1 \pm 1.3\,km/sec/Mpc$. The loss in gravitational energy will cause galaxies no longer to be co-moving and to move away from the Hubble flow. The increasing distances measured between them is known as dark energy.

5 A Cyclical Universe

The Friedmann equation postulates a perfect fluid to start. It will be valid only hours into the big bang when large shell fragments captured hot core gases forming proto-galaxies. The equations describing the scale factor evolution originate in the Ricci tensor. The 0-0 component of the Einstein equation is the Friedmann equation needing energy changes. In a Friedmann universe, radiation density is inversely related to the fourth power of the scale factor $\rho_r \propto a^{-4}$ and matter follows the third power $\rho_m \propto a^{-3}$.

$$\left(a_{,t}/a\right)^2 = -k/a^2 + \Lambda/3 + 8\pi\rho_m\, a_o^3/3a^3 + 8\pi\rho_r\, a_o^4/3a^4 \qquad (37)$$

Here a_o is our present day universe and Λ is Einstein's term for energy of empty space. The question has long been why the initial geometry of the universe was flat or $k = 0$ in the above equation. With the limiting matter density $\rho_m \approx 10^{17} grams/cm^3$, the universe was never radiation dominant. If all the initial radiation ρ_r was embedded in the matter and participated in the subsequent expansion and gravitation of a fairly large mass $r_{radius} \approx 10^{13}$ cm., then the flatness problem is explained. Due to the matching of gravitation and expansion energies ($k = 0$ above), it is most unlikely anything but nucleons stopped the prior universe collapse or restarted the expansion. There was no free radiation prior to the big bang. The Planck spectrum core photons participated in the re-expansion and nucleosynthesis. The small dark energy value attributed to Λ is actually due to later galactic black hole gravitational loss as explained. There is a question of increases of entropy on repeated universe cycles as irreversible physical processes may cause increasing cycles [24].

$$\frac{dS_0}{dt} = \frac{1}{T_0}\frac{dE_0}{dt} + \frac{p_0}{T_0}\frac{d(\delta v_0)}{dt} + \frac{\partial S_0}{\partial N_1}\frac{dN_1}{dt} + ... + \frac{\partial S_0}{\partial N_n}\frac{dN_n}{dt} \qquad (38)$$

As this is an adiabatic process in our universe, any entropy changes would be restricted to $N_1 \cdots N_n$. Since the universe goes through a black hole where

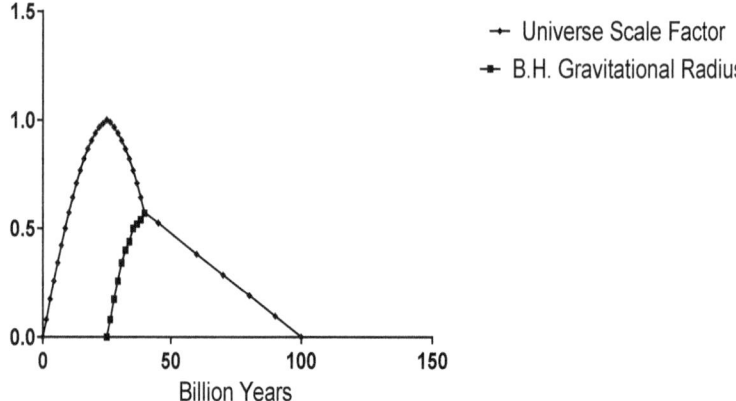

Figure 2: **The Cyclical Universe** started from a spherical shaped mass at or near limiting density. Due to the bounce, the Universe expands to a maximum and then contracts. During contraction, there is a growing ultra-max black hole. The collapsing universe will eventually force all matter and radiation into its gravitational radius. Kinetic energy of shell particles and radiation will slowly transfer to core. The gravitational energy and gravitational radius will slowly decrease to zero (cycle time estimated). Then core potential energy can start a new cycle of same size. Entropy is not increased.

only mass, electric charge and angular momentum count, any entropy gain would be lost when the highly squeezed nucleons are produced. Our universe cycles as follows as shown in Fig. 2. During the universe contraction phase, there is a growing ultra-max black hole. All matter and photons inside its growing gravitational radius will follow null geodesics into the inside matter. Their energy will be slowly transferred to highly squeezed core particles. The collapsing universe will eventually force all remaining matter and radiation into the growing r_+. With the loss of gravitational energy and entropy, a subsequent universe bounce will not increase in size from the previous cycle. By bounce time, the gravitational radius has decreased to zero, $r_+ \to 0$. The very energetic core nucleons, loaded with radiation, powered the re-expansion as well as resulting gravity and nucleosynthesis. Nucleosynthesis thus occurred only in the core, leaving the shell as cold dark matter. If all the matter in the universe were in one super-mass, its radius would be $\sim 10^{13} cm$. Its $4 - 5\%$ core released photons with an initial temperature $\sim 10^{13}$ degrees in a Planck spectrum. The cold baryonic shell surrounding the hot core absorbed and did not reflect hot core photons. It comprised a cavity close to the characteristic of a perfect black body with resulting radiation in thermal equilibrium.

Ihe initial mass expansion consisted of a cold shell surrounding the hot core. After thermal energy had been transferred to highly confined core baryons producing a Planck spectrum, gravitation loss permitted the expansion. The shell originated dark matter and black holes. Whether released from a small hole or the massive break up of the shell, subsequent light emission will have a Planck spectrum with temperature fluctuations $\sim 10^{-5}$. The energy density was

$$u_\nu(T) = \left(8\pi\nu^2/c^3\right)\left(h\nu/(e^{h\nu/k_B T} - 1)\right) \tag{39}$$

where k_B is Boltzmann's constant. The first term on the right represents the number of electromagnetic modes of the standing waves at frequency ν per volume of cavity. The second term represents the average energy per mode at this frequency. The primordial spectrum of curvature perturbations can be represented as a power spectrum

$$\mathcal{P}(k) \propto k^{n_s - 1} \tag{40}$$

WMAP[11] showed that the power spectrum $n_s = 0.963 \pm .014$, which is nearly scale free. The heights of the acoustic peaks are related to the related to the densities of the hot and cold (CDM) baryons. The bulk modulus is reduced by increasing baryon fraction which adds inertia but not pressure to the plasma. In the middle of the oscillations, the over-density increases the compression peaks $(1, 3, 5 \dots)$. The measured $\Omega_b h^2 = 0.02258 \pm .00056$ is consistent with nucleosynthesis. The angular scale of the acoustic peaks are related to r_s/d_a, which is the sound horizon co-moving distance divided by the angular distance back to the last scattering. This allows the spacetime curvature from the first acoustic peak to be measured. It was found to be flat to within 0.5%, all consistent with a bounce.

6 Baryonic Galaxy Formation

Big bang fireballs can not originate highly structured galaxies[5], [12], [13], [16], [20]. Like the little red dot galaxies found by the Webb telescope, black hole masses could have been present at the last scattering surface without enlarging the CBR isotropy.[27]. For most galaxies, there is a uniform history for galactic evolution[22]. There is a synchronization timescale ($T_s \approx 1.5$Gyr) where galaxies of fixed mass and red-shift go through a deterministic sequence of star formation, quasar accretion and eventual quiescence. This sequence negates the importance of stochastic processes. Galaxies can not form in a radiation dominant era so the very early galaxies found by the James Webb telescope are unexplained. The integrated Sachs-Wolf effect

$$\ell^2 C_\ell^{ISW} \simeq 72\pi^2/25\ell \int_0^{r_{LS}} dr\, r\, g'^2(r) \mathcal{P}_\mathcal{R}\Big(\ell/r\Big) T^2\Big(\ell/r\Big) \qquad (41)$$

was designed for photons descending into and emerging from a gravitational well. Here $\mathcal{P}_\mathcal{R}(k)$ is the primordial co-moving curvature spectrum. r_{LS} is the co-moving radius to the last scattering surface. g is the Λ growth suppression factor. $T(k)$ is the transfer function for suppression during radiation domination. k' is the conformal time derivative of the co-moving wave number. Black hole masses greater $10^4 M_\odot$ surrounded by hot gasses could have been present in the last scattering surface without enlarging the isotropy from $10{-}^5$. Starting from a big bang shell and hot core, early galaxy formation and high inter-galactic correlations become simplified. The shell laid down a fairly even cold dark matter density ρ_{DM}. The DM halo mass distribution for galactic systems ranging from dwarf discs and spheroidals to spirals and ellipticals has been found essentially constant[7]. This result also spans almost the whole galaxy magnitude range M_B from -8 to -22 and gaseous to stellar mass fraction range of many orders of magnitude.

$$log(\mu_{0D}/M_\odot pc^{-2}) = 2.15 \pm 0.25 \qquad (42)$$

where μ_{0D} is the central surface density and is defined as $r_0\rho_0$. r_0 is the halo core radius and ρ_0 is the central density. This same finding was supported by another group [9]. Since all dark matter lies within a halo orbiting the primordial black holes, its total density field is

$$\rho(\mathbf{x}) = \Sigma_i \int dm \int d^3x \prime \delta(m - m_i)\delta(\mathbf{x}\prime - \mathbf{x}_i) m\, u(\mathbf{x} - \mathbf{x}\prime|m) \qquad (43)$$

132

where i is the different halos, $u = \rho/M$ is the normalized density profile and M is the halo virial mass. Within the original dark matter halos, massive black holes coalesced. The dark matter from the center to periphery of early type galaxies has been evaluated from a galactic stellar mass $M_* \sim 10^{10} M_\odot$ to the more massive galaxies $10^{12} M_\odot$[26]. N-body simulations predict the the dark matter density profile $\rho_{DM}(r)$ should be independent of halo mass. In the NFW profile it is described by two power laws. In the outer regions it is $\rho_{DM}(r) \propto r^{-3}$ and in the center $\rho_{DM}(r) \propto r^\alpha$. Here α can vary -1 or -1.5 depending on the model. What has actually been found is a variation around the inverse gravitational square law ($\alpha = -2$), as the halos are orbiting the primordial black holes. Galaxies larger than $3 \times 10^{10} M_*$ have lower slopes than $\alpha \approx 2$ due to accretion of halo mass from smaller satellite galaxies. Smaller satellite galaxies under $3 \times 10^{10} M_*$ have larger slopes due to loss of outlying halo matter. The rotation of the larger galaxies above this break point has been found highly correlated and perpendicular to the filament that they are located [8]. The rotation of 65 galactic black holes has been found aligned using their radio galactic jets [23].

At first large and small shell masses were driven out forming filaments. Larger masses coalesced into black holes which held a fairly even density of smaller dark matter. Later hot core gas was captured according to the depth of the gravitational well. During capturing process, larger black holes, unlike smaller black holes, did not change the direction of rotation. The deeper the gravitational wells, the higher the velocity and more orbiting mass that could be captured. The capturing process described is divided by distance from the primordial black holes M. Outside the immediate area of black hole influence, capturing of hot core matter m streaming through the area of influence of each black hole is due to the amount of energy each particle possesses. Large kinetic energies result in hyperbolic or parabolic type orbits with the ability to escape any given gravitational well. Matter that is captured has the potential energy greater than the kinetic,

$$GmM/r > l^2/mr^2 + 1/2\, m\dot{r}^2 \tag{44}$$

and $e < 1$. Expanding the total kinetic energy E in the equation for e,

$$e = \left[1 + (2l^2(l^2/mr^2 + 1/2\, m\dot{r}^2 - GmM/r))/mk^2\right]^{0.5} \tag{45}$$

Orbiting matter has $e < 1$ and real. If we let its angular momentum $l = mr\dot{\theta}^2$

and $k = mMG$, the equation for e becomes

$$e = \left[1 + (r^6\dot{\theta}^4 + \dot{r}^2 r^4 \dot{\theta}^2 - 2GMr^3\dot{\theta}^2)/(M^2G^2)\right]^{0.5} \tag{46}$$

Using $\dot{\theta} = \dot{r}/r$, the equation for e becomes

$$e = \left[1 + (2r^2\dot{r}^4)/(M^2G^2) - (2r\dot{r}^2)/(MG)\right]^{0.5} \tag{47}$$

As $GM = \dot{r}^2 r$, then the galactic well will deepen as $M \propto \dot{r}^2$ or $M \propto r$. The last term in equation above becomes $\dot{r}^8/M^2G^2$. When this term is dominant, it will allow capturing matter with $\dot{r}$ to increase as the fourth power as the galactic black hole M increases, $\dot{r} \propto M^4$. This explains the Tully-Fisher and similar correlations[21]. The black hole capturing cross sectional area, $M_{csa} \propto M_{gravity}$ since both scale as r^2.

The two stage gravitational formation process preserves angular momentum from the big bang.. Halo parameters are related to the luminous mass distribution since all rotating mass was captured by a given size black hole. An entirely baryonic model explains why the circular orbital speed from luminous matter, which dominates the inner regions, is so similar to dark matter at larger radii. With many stars in the center areas, initial conditions for dark and luminous matter no longer have to be closely adjusted to produce a flat rotation curve[10]. The hot core matter of a certain velocity can be captured by the similarly sized black holes, explaining why there are similar circular speeds in all galaxies of a given luminosity no matter how the luminous matter is spaced. The depth of the gravitational well determines the circular speed and luminosity of captured matter. The hot and cold matter discrepancies are detectable only at accelerations below $\sim 10^{-8} cm/sec^2$ since they are all baryons. Much of the missing baryonic matter has been found in the intergalactic medium[15].

7   Discussion

General relativity has been extrapolated in black holes and the big bang to enormous energies and densities without consideration of properties of matter. Lightly squeezing nucleons like neutron stars will result in a light reduction of gravitational energy. Highly squeezing nucleons as in pre-big bang matter and black holes will result in elimination of gravitational energy.

Highly squeezing matter will cause quantum effects in the stress-energy tensor so that $T_{\mu\nu} \to \mathbf{0}$. Very simple 'quantum gravity' can be produced. The gravitational losses including dark energy can't be explained any other way. Once damping of motion in highly squeezed particles is included, general relativity remains applicable throughout the Universe. A cold shell and hot core produced the initial Planck spectrum radiation. The cold shell dispersion led to the great wall, filaments and voids. The basis of almost all galaxies formed simultaneously as cold dark shell matter coalesced into black holes. which captured subsequent hot core gases into proto-galaxies. The formed stars ionized the intergalactic medium. The fact that the Higgs Boson is 125 GeV and not larger, eliminates nonbaryonic matter as dark matter.

Gravitation is described by the Riemann metric. No negative energies are required nor is a vacuum energy necessary for empty space. The fact that initial universe had flat space-time and a Planck spectrum has been explained. Since the origin of the big bang is a bounce, it should re-collapse into the big crunch. The resulting neutrons will eventually decompose into the protons and electrons to make hydrogen and helium for the next cycle. Black holes must continuously accrete mass-energy to maintain their gravitational strength. Gravity is not caused by matter itself but rather by the motion of matter particles. Galaxies were made in a two stage process. First the big bang shell laid down all halos and super-massive black holes. Then hot core gases were captured according to the size of each gravitational well.

References

[1] Almheiri, A., Marlof, D., Polchinski, J., & Sully, J. Black Holes: Complementary or Firewalls? Preprint at arXiv.org:1207.3123, (2012).

[2] Bally, J. 2015, Mystery survivor of a supermassive black hole. *Nature* **524**, 301-302, (2015).

[3] Bianchi, S. et al. *HST unveils a compact mildly relativistic Broad Line Region in the candidate true type 2 NGC 3147* Preprint at arXiv.org:1905.09627, (2019).

[4] Burkert, V.D., Elouadrhiri, L & Girod, F.X. The pressure distribution inside the proton *Nature* **557**, 396-399, (2018).

[5] Cattaneo, A. et al. The Role of Black Holes in Galaxy Formation and Evolution. *Nature* **460**, 213-219, (2009).

[6] Dallilar, Y. et al. A Precise measurement of the magnetic field in the corona of black hole V404 Cygni. *Science* **358**, 1299-1302, (2017).

[7] Donato, F. et al. A Constant Dark Matter Halo Surface Densities in Galaxies. *Mon. Not. R. Astron. Soc.* **397**, 1169-1176,(2009).

[8] Dubois, Y. et al. Dancing in the dark: galactic properties trace spin swings along the cosmic web. Preprint at arXiv.org:1402.1165, (2014).

[9] Gentile, G., Famaey, B., Zhao, H. & Salucci, P. Universality of galactic surface densities within one dark matter halo scale-length. *Nature* **461**, 627-628, (2009).

[10] Ibata, R.A. et al. A Vast Thin Plane of Co-rotating Dwarf Galaxies Orbiting the Andromeda Galaxy. *Nature* **493**, 62-65, (2013).

[11] Komatsu, E. et al. Seven Year Wilkinson Microwave Anisotropy Probe: Cosmological Observations. *Astrophys. J. Suppl.* **192**, 18, (2011).

[12] Kormendy, J., Bender, R., & Cornell, M.E. Super-massive Black Holes Do Not Correlate with Galaxy Disks or Pseudo-bulges *Nature* **469**, 374-376, (2011).

[13] Lerner, E. Obervations contradict galaxy size and surface brightness predictions that are based on the expanding universe hypothesis. Mon. Not. Royal Astron. Soc. (2017).

[14] Misner, C., Thorne, K.S., & Wheeler, J.A. *Gravitation* (W.H. Freeman and Co., New York 1973) p131-151, 627, 846-871.

[15] Nicastro, F. et al. Observations of the missing baryons in the warm-hot intergalactic medium. *Nature* **558**, 375-376, (2018).

[16] Peebles, P.J.E. How Galaxies Got Their Black Holes. *Nature* **469**, 305-306, (2011).

[17] Perlmutter S. et al.Discovery of a Supernova Explosion at half the age of the Universe and its Cosmological Implication. Preprint at arXiv.org:astroph/97122. 1997

[18] Ponti,G., De, K., Munoz-Darias, T., et al. The puzzling orbital period evolution of the LMXB AXJ1745.6-2091, Preprint at arXiv.org:1511.02855, (2015).

[19] Riess AG et al Observational Evidence from supernovae for an accelerating universe and a cosmological constant. Preprint at arXiv:astro-ph/9805291. 1998

[20] Rosenberg DE. Biblical Genesis vs. Science's Big Bang. Pittsburgh: Dorrance Publishing; 2020 pp. 73-76, 83-88

[21] Shaya E.J. & Tully R.B. The Formation of the Local Group Planes of Galaxies *Mon. Not. R. Astron. Soc.* **436**, 2096-2119 (2013).

[22] Steinhardt, C.L. & Speagle, J.S. A Uniform History of Galaxy Evolution. Preprint at arXiv.org:1409.2883, (2014).

[23] Taylor, A.R. & Jagannathan, P. Alignments of radio galaxies in deep radio imaging of ELAIS N1. *Mon. Not. R. Astron. Soc.* **459**, L36-L40, (2016).

[24] Tolman, R. Possibilities in Relativistic Thermodynamics for Irreversible Processes Without Exhaustion of Free Energy *Physical Review* **39**, 329-346, (1931).

[25] Thorne, K. Nonspherical Gravitational Collapse: Does it produce Black Holes? *Comments on Astrophysics and Space Physics* **2**, 191-196, (1970).

[26] Tortora, C. et al. Systematic variations in central mass density slopes in early type galaxies. Preprint at arXiv.org:1409.0538, (2014).

[27] Vandenbergh, S. How Do Galaxies Form? *Nature* **455**, 1049-1051, (2008).

[28] Will, C.M. On the unreasonable effectiveness of the post-Newtonian approximation in gravitational physics. *Proc. Nat. Acad. Sci. (US)* **108**, 5938, (2011).

ADDITIONAL REFERENCES

Brehm, J. J. & Mullin, W.J. Introduction To The Structure Of Matter (John Wiley and Sons, New York) 73-99, (1989).

Gutfreund, H. & Renn,J. The Road to Relativity (Princeton University Press, Princton). (2015)

Lee, H.J. & Schiesser, W.E. Ordinary and Partial Differential Equations Routines... (Chapman & Hall/CRC, Boca Raton) (2004)

Loeb, A. & Furlanetto, S.R. The First Galaxies in the Universe (Princeton University Press, Princeton 2013).

Mahon, B. The Forgotten Genius of Oliver Heaviside(Promethius Books, Amherst NY (2017)

Misner, C., Thorne, K.S., & Wheeler, J.A. Gravitation (W.H. Freeman and Co., New York 1973) p131-151, 627, 846-871.

Susskin, L. The Black Hole War (Little, Brown and Co., New York 2008)
Acknowledgments This article is based on research I posted on the internet astrophysics archives while in NJIT Physics. Some of it was done in collaboration with John Rollino of Rutgers University Physics, Newark N.J. 07102 Unfortunately Prof. John Rollino has since passed away. I wants to thank Erik Schnetter for helpful advice.